Topics in Applied Physics Volume 22

Topics in Applied Physics Founded by Helmut K. V. Lotsch

X-Ray Optics

Applications to Solids

Edited by H.-J. Queisser

With Contributions by
A. Authier   U. Bonse   R. Feder   W. Graeff
W. Hartmann   S. Kozaki   H.-J. Queisser
E. Spiller   M. Yoshimatsu

With 133 Figures

Springer-Verlag Berlin Heidelberg GmbH 1977

Professor Dr. *Hans-Joachim Queisser*

Max-Planck-Institut für Festkörperforschung, Büsnauer Straße 171
D-7000 Stuttgart 80, Fed. Rep. of Germany

ISBN 978-3-662-30913-1   ISBN 978-3-540-37042-0 (eBook)
DOI 10.1007/978-3-540-37042-0

Library of Congress Cataloging in Publication Data. Main entry under title: X-ray optics. (Topics in applied physics; v. 22). Bibliography: p. Includes index. 1. Solids—Optical properties. 2. X-rays. 3. Solid state physics. I. Queisser, Hans-J., 1931-. II. Authier, A. QC 176.8.06X2   530.4'1   77-15086

This work is subject to copyright. All rights are reserved, whether the whole or part of the material is concerned, specifically those of translation, reprinting, re-use of illustrations, broadcasting, reproduction by photocopying machine or similar means, and storage in data banks. Under § 54 of the German Copyright Law, where copies are made for other than private use, a fee is payable to the publisher, the amount of the fee to be determined by agreement with the publisher.

© by Springer-Verlag Berlin Heidelberg 1977

Originally published by Springer-Verlag Berlin Heidelberg New York in 1977
Softcover reprint of the hardcover 1st edition 1977

The use of registered names, trademarks, etc. in this publication does not imply, even in the absence of a specific statement, that such names are exempt from the relevant protective laws and regulations and therefore free general use.

2153/3130-543210

Preface

Optics has been rejuvenated during the last few decades, mostly through the discovery of the laser. Most activity is, however, confined to the spectral regions of the visible radiation and adjoining regimes. The X-ray region, on the other hand, experienced a different development. Plasma diagnostics and the related field of X-ray astronomy provided many stimuli for improved instrumentation and novel experiments. Medical and materials-testing applications, of course, continued to provide strong motivation for the improvement of sources, detectors, and signal processing techniques in the X-ray region. The availability of synchrotrons and storage rings as powerful sources for X-rays also had a significant impact on utilization of X-ray photons for a variety of experiments in many fields.

This volume cannot cover all these aspects. The necessary restriction of topics therefore emphasizes those new developments in X-ray optics which are related to solid-state sciences and particularly to new solid-state technologies. The advancements of solid-state physics, the chemistry and crystallography of solids, and the understanding and control of X-radiation have always been closely intertwined and will continue to be intimately coupled to each other. One example is given by modern semiconductor technology with its achievement of highly perfect crystals and its requirements for extremely stringent control of crystalline perfection as well as with its need for geometry definition beyond that achievable with visible light. The contributions of this volume have been selected with this interrelation of X-ray optics and solid-state sciences in mind. The well-known techniques of crystallographic structure determinations are not covered here. Instead, it was attempted to accentuate new approaches, many of which we presume to emerge from the research laboratories into widespread applications for future fabrication of solid-state materials and devices in a similar fashion as numerous other applications of X-rays have done previously.

The modern status of high-power X-ray sources is covered first; synchrotron radiation is not treated in this book since an entire separate volume of the series "Topics in Current Physics" (Editor: C. Kunz) will later be devoted to this topic. Chapter 3 of the present volume covers in great detail the scientific principles and the practical applications of X-ray lithography, an extremely rapidly moving field of vital importance for the future of microminiaturization of semiconductor devices and other highly integrated solid-state structures. Chapter 4 reviews the present state of interferometry and includes the inseparable field of neutron interferometry as well. The two final chapters concern themselves with modern facets of the technique of X-ray topography, which is a large-area mapping of almost perfect single crystals in the light of specific Laue reflections. Section topography, treated in Chapter 5, is a particularly sensitive tool to use for revealing the fine structure of defects and obtaining information on depth distributions of such defects; live topography, as described in Chapter 6, is the attempt to avoid lengthy photographic exposure by using a direct display technique.

I hope to have chosen a coherent, closely interrelated set of subjects out of the vast field of modern X-ray techniques, which may be of use to those interested in applying and advancing both X-ray technology and solid-state sciences. I would like to thank all those who have helped me in the editing of this book—first all the authors for their cooperation in trying to present the most modern status of their subject fields; then the publisher for the speedy publication, W. Hagen for his help and advice, and Miss A. Vierhaus for her patience and skill in typing and handling.

Stuttgart, August 1977   *Hans-Joachim Queisser*

Contents

Contributors

Authier, André
  Laboratoire de Minéralogie et Cristallographie, Université de Paris VI,
  Tour 16, 4 Place Jussieu,
  F-75230 Paris Cedex 05, France

Bonse, Ulrich
  Institut für Physik, Universität Dortmund, Postfach 500500
  D-4600 Dortmund 50, Fed. Rep. of Germany

Feder, Ralph
  Thomas J. Watson Research Center, IBM Corporation,
  Yorktown Heights, New York, NY 10598, USA

Graeff, Walter
  Institut für Physik, Universität Dortmund, Postfach 500500
  D-4600 Dortmund 50, Fed. Rep. of Germany and
  Institute Laue-Langevin, F-38042 Grenoble, Cedex, France

Hartmann, Werner
  Max-Planck-Institut für Festkörperforschung, Büsnauer Straße 171,
  D-7000 Stuttgart 80, Fed. Rep. of Germany

Kozaki, Shigeru
  Rigaku-Denki Company Ltd., Haijima, Akishima, Tokyo 196, Japan

Queisser, Hans-Joachim
  Max-Planck-Institut für Festkörperforschung, Büsnauer Straße 171,
  D-7000 Stuttgart 80, Fed. Rep. of Germany

Spiller, Eberhard
  Thomas J. Watson Research Center, IBM Corporation,
  Yorktown Heights, New York, NY 10598, USA

Yoshimatsu, Mitsuru
  Rigaku Denki Company Ltd., Haijima Akishima, Tokyo 196, Japan

1. Introduction: Structure and Structuring of Solids

H.-J. Queisser

1.1 Scope

This volume combines several articles concerning novel techniques, in which X-ray optical methods are utilized for the analysis and processing of solid-state materials. X-ray techniques have consistently been vital tools for the understanding of crystalline solids and their ideal as well as real atomic arrangements [1.1, 2]. The necessary restriction for this volume led to a selection of new techniques which presently appear particularly promising for playing a significant role in the future development of advanced mastering of solids, such as in the microelectronics of large-scale integration. Modern X-ray methods lead to increased understanding of the crystal structure as well as to highly resolved compositional and geometrical structuring of solids [1.3].

Emphasis is therefore first placed on refined techniques in which X-ray optics can reveal lattice defects in single crystals of high perfection. Here, silicon and quartz are the typical prototype materials. Demonstration of new X-ray methods for fine structure analysis of well-defined isolated imperfections has emerged in parallel with the cultivation of large, highly controlled single crystals. The two contributions on X-ray topography in this volume deal with this aspect. X-ray interferometry, another topic of this book, is a new experimental method in which strong mutual interaction with material science exists also. The demonstration of the feasibility of X-ray interferometers led to extension into the neutron regime, but also enabled some basic experiments in the physics of elementary particles to be performed.

Optical technology in the visible spectral regime has played a significant part in the rise of new generations of microelectronic devices, especially in the field of semiconductor circuits of ever-increasing number and density of functions. The wavelength of the commonly employed ultraviolet spectral range now limits further substantial increases in density and complexity [1.4]. Shorter wavelengths can be provided for enhanced spatial definition by either electron beams or by X-rays. One chapter of this book therefore deals with the rapidly developing field of X-ray lithography.

Sources of X-rays have been worrisome because of the notoriously low conversion efficiency of electrical input power into X-ray photons [1.5]. This feature had limiting effects on the ease and applicability of X-ray methods. The first chapter of this book therefore deals with the strongest conventional sources which are available today: rotary anode generators. Most of the methods described later presently rely on such generators.

Another type of source is currently becoming available; it promises to lead to many important results. This source is "synchrotron radiation", originating from accelerated electrons in synchrotrons or storage rings [1.6]. This very efficient contemporary development is not covered by a chapter here since an entire volume of the series "Topics in Applied Physics" is planned for the near future.

Missing in this volume are many other subjects falling into the category of X-ray techniques related to solid-state science and technology. Space restriction forces us to merely list some of them here and stimulate interest by quoting a few references, obviously without any claim of completeness. The entire field of crystal structure determination [1.1, 7], especially with the new developments in computer-automated data processing [1.8], is not represented in this volume. Diffuse scattering analysis [1.9] and X-ray Raman scattering [1.10] could not be covered here. Fluorescent analysis [1.11], stress analysis [1.12], and the vast, rapidly expanding field of X-ray photoelectron spectroscopy (ESCA) [1.3] cannot be touched upon here. There exists a great deal of literature on these topics; some fields of particular significance are currently under consideration for the present series "Topics in Applied Physics". Soft X-ray spectroscopy, for example using grating spectrometers, is adequately described elsewhere [1.14]; X-ray microscopy has seen a number of new developments since the publication of the book by *Cosslett* and *Nixon* [1.15]. The need for imaging of X-ray emitting objects for a subsequent spatial or spectral analysis has been particularly important for plasmas, such as in astrophysics or for fusion research [1.16].

1.2 X-Rays and Solids

1.2.1 Optics and Solids

The remarkable renaissance of optics during the last two decades was in part due to progress in solid-state materials and devices. The most noteworthy examples are the ruby laser, semiconductor injection lasers, light-emitting diodes, and the great variety of solid-state radiation detectors as well as the low-loss fiber optical waveguide materials. Controlled and efficient mutual conversion between electromagnetic radiation and electronic processes in the solid is now feasible in the visible and infrared regimes. The injection laser [1.17] demonstrates that compositional control within fractions of a micron inside a semiconductor is a prerequisite to obtain a source for coherent radiation in the wavelength range of about one micron. Comparable accuracies and utilizations in the ultraviolet range and beyond are, however, still seriously impeded by the stringent conditions for control of very small dimensions and the onset of high absorption. Solid-state technology has begun to change optics markedly: traditional free-space propagation with utilization of individual optical components is gradually being replaced by a technique of producing, guiding, modulating, and converting electromagnetic radiation within highly structured solids, the aim of this development being a truly "integrated optics" [1.18].

1.2.2 X-Rays and Solid-State Physics

Solids have played a decisive role in the history of X-ray physics and technology. Experimental proof of the wave nature of X-rays was achieved in the classical fashion by diffraction at slits; however, of revolutionary character was the diffraction experiment using the three-dimensional grating of a regular single-crystal solid. This experiment by *Laue* and his collaborators [1.19] immediately opened up the field of diffraction analysis for solids, thus marking the beginning of solid-state physics with the experimental proof of the essential periodicity of a solid and the precise determination of interatomic distances. A direct result was the early description of a solid within reciprocal space, a concept which later became an essential foundation of all solid-state theory. Later developments focused on the real structure of solids, in particular the extremely strong influence of specific lattice defects, which could be positively identified and quantified only by X-ray or electron diffraction techniques.

The understanding of wave fields within a solid of lattice constant about equal to the wavelength was achieved through the dynamical theory [1.20] and the interpretation of such unusual phenomena as the anomalous transmission [1.21]. Crystallography and wave mechanics had to be combined when the characteristic dimensions, lattice parameter and wavelength, become of equal magnitude.

1.2.3 Outlook

The intricate and complex effects resulting from this interaction between an electromagnetic wave field and a regular atomic array still represent a fruitful area of research, development, and applications. It is my feeling that we can presently expect a time of renewed achievements in this field. I base this hope on the vigorous efforts and remarkable success already apparent in the control of the composition of solids. One particularly striking example has been given through the techniques of molecular-beam epitaxial crystal growth [1.22]: it now appears feasible to control individual layers of a mixed-crystal semiconductor material [1.23]. The techniques of proving and analyzing this remarkable claim of material control down to atomic dimensions must obviously be based on diffraction techniques. On the other hand, it now seems realistic to anticipate that solid structures of such finely tuned character with designed lattice parameter variations or having prescribed atomic species within specified lattice planes cannot fail to make their impact in confining, guiding, or modulating wave fields in the X-regime to heretofore unknown precision. New effects and applications will thus emerge from these — admittedly difficult and painstaking — experiments in designing solids. It is the purpose of this volume to contribute towards progress in this interdisciplinary task by combining solid-state aspects with X-ray technology.

1.3 Sources

The standard arrangement of any optical experiment consists of the sequence: source, propagation medium, specimen, detector. For each spectral region usually one of these components clearly limits experimental efficiency and accuracy. For example, in the infrared, the detector problem presents particular difficulties due to the small quantum energy of those photons. In the X-ray regime, however, the photons are sufficiently energetic to utilize a wide range of reactions and effects for detection, although noise may present problems. (A discussion of the detector aspect will be included in Chap. 6, since live topography cannot be satisfied with integration in a photographic plate but is forced to find an efficient down-conversion process into visible photons.) The source is therefore the central experimental problem.

X-ray sources typically have conversion efficiencies of only fractions of a percent and are beset by the task of rapid removal of the excess heat which would otherwise damage the active anode region. Today's very widely used sources are therefore of the rotary anode type. A great deal of development work has gone into these generators. Chapter 2 of this book, by *Yoshimatsu* and *Kozaki*, is a detailed description of the history, the present state of the art, and the anticipated trends in this technology.

X-ray lasers are today the subject of much discussion and speculation [1.24]. There is currently a great deal of activity in producing and verifying a coherent emission, usually in the region of very soft X-rays having wavelengths of several hundreds of Angstroms. We felt that this field had not yet sufficiently matured at the time of writing this book to warrant inclusion of this topic. It is, however, quite obvious that any breakthrough in obtaining a laser-like source for X-rays would greatly influence and stimulate many applications, including those covered by the chapters of this volume [1.27].

1.4 Structuring with X-Ray Techniques

Spiller and *Feder* cover a topic of great current interest in Chapter 3 of this volume. X-ray lithography for solid-state device applications is being considered as a feasible way to greatly enhance the degree of geometry control which is an absolute necessity for increasing the density of integrated circuit functions and reducing the size of their individual elements. The idea of using wavelengths shorter than the usual visible light to improve microscope resolution is, of course, an obvious one and has been treated in great detail for quite some time [1.15]. It appears, however, that the definite industrial need for finer structuring and the associated great effort in this field will lead to renewal of interest in this general field of X-ray microscopy with far-reaching applications in many other fields, including, of course, biology and medicine. An example of this is the work by *Spiller* et al. [1.25], in which they used the subsequent inspection of their lithographic image by means of a highly magnifying scanning electron microscope to achieve extremely high resolution. The future will show whether the further development of such combined techniques might eventually lead to an optimal microscopy, to be applied to biological macromolecules.

The lithographic activity of the impinging X-rays results from the liberation of photoelectrons in a suitable resist, which then leads to sharply localized development of this photosensitive resist with ensuing usable changes in chemical features, such as solubility or etching rates. The question arises why there is need to first convert high-energy electrons into X-ray radiation if it is the reconverted electrons that cause the essential effect. Indeed, development of electron beam lithography is currently receiving greater emphasis than is X-ray lithography for solid-state device applications. One can often hear the opinion today that the rapidly progressing technology of electron beam lithography may render X-ray techniques uninteresting. There are, however, advantages for X-ray methods. The conversion to electromagnetic radiation facilitates propagation, especially through interfaces such as windows. The reduced sensitivity to external fields favors waves over charged particles. Electron scattering and charging effects are avoided. There is less sensitivity to dust and contamination in the exposure process. The essential problem of aligning the specimen for multiple consecutive exposures requires attention for X-ray systems, especially using synchrotron radiation. All these options are, however, as yet not thoroughly determined and will be strongly influenced by considerations of economy.

1.5 Interferometry

Optics in the X-ray regime is characterized by a lack of dispersive effects, on which elements such as lenses or prisma are based in other spectral regions. The external frequencies greatly exceed the frequencies of the polarization oscillators. One can therefore approximate the index of refraction in the X-ray region by

$$n = 1 - C\varrho\lambda^2 \tag{1.1}$$

where ϱ is the density of the material and λ the wavelength. The constant C is of the order of 10^{11} m kg^{-1}. This equation shows the independence of specific material properties. We further see that the index n is always slightly smaller than unity. For a density of $10 \, \text{g cm}^{-3}$, one has at $1 \, \text{Å}$ merely a deviation of about 10^{-5} from unity. Utilization of this small difference of the index of refraction versus vacuum would lead to impractically long focal lengths.

Bragg reflection from regularly spaced atomic planes of a solid can fortunately be utilized for optical components. A very sophisticated type of interferometry, using highly perfect single crystals, has recently arisen. *Bonse*, one of the original researchers in this new field, and *Graeff* describe the principles and applications of these methods in Chapter 4. X-ray radiation is so similar here to neutrons of comparable wavelengths that these two types of undular excitation have been combined. Neutron interferometry has permitted a very beautiful experiment of elementary nature, in which the phase shift caused by gravitational interaction could be determined; this experiment is the only one where the constant h of quantum theory appears together with the elementary constant of gravitation.

X-ray interferometry has been applied to optical comparison of materials against a standard of the same material. The extreme conditions of geometry control require that the individual components of the interferometer must often be machined out of one perfect crystal with long crystallographic coherence. This requirement has thus far somewhat restricted a more general usage of X-ray interferometry for, say, silicon materials characterization. Hopefully, further progress will occur, leading to a novel, highly sensitive probe for crystalline structure which ought to be useful for the undoubtedly yet-increasing standards of crystalline perfection for future solid-state materials.

1.6 Defect Structure Topography

Deviations in the regular crystallographic arrangement are caused by lattice defects, such as dislocations, stacking faults, and internal boundaries. These defects may cause uncontrolled alterations in many physical parameters and thus present undesirable properties of the material. Diffraction techniques are well suited for the detailed analysis of such defects; X-ray diffraction is specifically advantageous since it is a nondestructive technique not requiring special sample preparation. The crystal may be viewed in the reflected light of one particular Laue reflex. The information concerning the crystalline perfection is contained in the intensity variations of this reflex. A complete image of the defect structure for a large sample can then be obtained by systematic scanning of the specimen and provision of a definite correlation between sample and detector, for example by rigidly connecting them and moving the entire arrangement with respect to the incoming X-ray beam. This technique, called topography, is of great practical importance today for the characterization of materials as well as for investigating the influence of processing on crystal structure [1.26].

This volume does not concern itself with a general review of topography, but describes two specific trends of topography in more detail. *Authier* treats in Chapter 5 the technique of section topography. This method yields detailed information on the nature of defects in nearly perfect crystals and enables one to make specific statements concerning the depth distribution of defects in comparatively thick samples. The predominantly theoretical nature of *Authier*'s contribution indicates the great amount of mathematical treatment necessary for extracting structural information from these disturbances of the wave fields within the crystal.

Topography today uses mostly photographic emulsions for mapping the image of a specimen. This restriction is caused by the comparatively weak sources, calling for integration of the X-ray flux by the photographic plate. A nonmagnifying technique results, which in turn places high requirements on the quality and grain size of the plate in order to retain detailed information. In situ measurements of dynamic effects are impossible. Topography has thus resisted the general trend of replacing analog detection methods with principles of digital counting. Coupling to modern signal processing methods is absent here, which represents a definite weakness of the method. It is therefore understandable that considerable efforts have been applied to finding a direct viewing technique, which is usually called "live topography". A set of digital signals which can be further processed by using the readily available techniques of television electronics and digital data handling leads to the goal of all of these techniques—a final television image. *Hartmann* summarizes in Chapter 6 the present state of this development which promises not only to furnish more detailed information for the technology of materials processing but also gives hope for a better basic understanding of defect dynamics.

References

1.1   A comprehensive collection of articles describing progress in the general field of X-rays is: *Advances in X-Ray Analysis*, ed. by C. S. Barrett et al. (Plenum Press, New York, each year one volume, representing the Proceedings of the Annual Conference on Applications of X-ray Analysis).

1.2   The situation up to 1963 is summarized in: *X-Ray Optics and X-Ray Microanalysis*, ed. by H. H. Pattee, V. E. Cosslett, A. Engström (Academic Press, New York 1963)

1.3   *Modern Diffraction and Imaging Techniques in Material Science*, ed. by S. Amelinckx, R. Gevers, G. Remaut, J. van Landuyt (North Holland, Amsterdam 1970)

1.4   A. N. Broers: "Recent Lithography Trends", in *Proceedings of the Symposium on Electron and Ion Beam Technology*, 7th Intern. Conf., ed. by R. Bakish (The Electrochemical Society, Princeton 1976)

1.5   W. Schaaffs: "Erzeugung von Röntgenstrahlen", in *Handbuch der Physik*, ed. by S. Flügge, Vol. XXX (Springer, Berlin, Heidelberg, New York 1957)

1.6   M. L. Perlman, E. M. Rowe, R. E. Watson: Phys. Today **27**, 30 (1974)

1.7   M. J. Buerger: *Contemporary Crystallography* (McGraw-Hill, New York 1970)

1.8   See, for example, W. Parrish, T. C. Huang: Acta Cryst. A **31**, 197 (1975)

1.9   W. A. Wooster: *Diffuse X-Ray Reflections from Crystals* (Clarendon Press, Oxford 1962)

1.10 P. Eisenberger, P. M. Platzman, H. Winick: Phys. Rev. Lett. **36**, 623 (1976)

1.11 L. S. Birks: *X-Ray Spectrochemical Analysis* (Interscience, New York 1959)

1.12 See, for example, B. D. Cullity: *Elements of X-Ray Diffraction* (Addison-Wesley, Reading, Mass. 1976) Chap. 17

1.13 T. A. Carlson: *Photoelectron and Auger Electron Spectroscopy* (Plenum Press, New York 1976)

1.14 In *X-Ray Spectroscopy*, ed. by L. V. Azároff (McGraw-Hill, New York 1974)

1.15 V. E. Cosslett, W. C. Nixon: *X-Ray Microscopy* (Cambridge University Press, London 1960)

1.16 F. Seward, J. Dent, M. Boyle, L. Koppel, T. Harper, P. Stoering, A. Toor: Rev. Sci. Instr. **47**, 464 (1976)

1.17 M. B. Panish, I. Hayashi: "Heterostructure Junction Lasers", in *Appl. Solid State Science*, ed. by R. Wolfe, Vol. 4 (Academic Press, New York 1974) p. 235

1.18 T. Tamir (ed.): *Integrated Optics*, Topics in Applied Physics, Vol. 7 (Springer, Berlin, Heidelberg, New York 1975)

1.19 M. Laue, W. Friedrich, P. Knipping: Ann. Physik **41**, 971 (1913)

1.20 M. v. Laue: *Röntgenstrahlinterferenzen*, 3rd. ed. (Akademische Verlagsgesellschaft, Frankfurt 1960)

1.21 G. Borrmann: Z. Physik **42**, 157 (1941)

1.22 A. Y. Cho, J. R. Arthur, Jr.: "Molecular Beam Epitaxy", in *Progress in Solid State Chemistry*, ed. by J. McCaldin, G. Somorjai, Vol. 10 (Pergamon Press, New York 1975) pp. 157

1.23 A. C. Gossard, P. M. Petroff, W. Wiegman, R. Dingle, A. Savage: Appl. Phys. Letters **29**, 323 (1976)

1.24 G. Chapline, L. Wood: Phys. Today **28**, 40 (1975)

1.25 E. Spiller, R. Feder, J. Topalian, D. Eastman, W. Gudat, D. Sayre: Science **191**, 1172 (1976)

1.26 B. K. Tanner: *X-Ray Diffraction Topography* (Pergamon Press, New York 1976)

1.27 A. Yariv and P. Yeh: Opt. Commun. **22**, 5 (1977)

2. High Brillance X-Ray Sources

M. Yoshimatsu and S. Kozaki

With 15 Figures

2.1 History of the Development

Ever since *Laue* established the undulatory nature of X-rays and *Bragg* analyzed the structure of ionic crystals by using X-ray diffraction, the X-ray diffraction technique has been developed as an important tool for clarifying varied properties of substances. In recent years, the application of X-ray diffraction has been further extended, as in X-ray diffraction topography, to observe lattice defects in crystals.

X-ray tubes used for diffraction today are based on *Coolidge*'s principle. Their construction is such that the cathode and the anode are opposed to each other over a short distance in a high vacuum. The cathode is provided with a tungsten filament. By heating the filament and varying its temperature, the number of thermal electrons which are emitted can be controlled independently of the potential gradient between both electrodes. In this way, *Coolidge*'s method has made possible the stable regulation of the intensity of X-rays emitted from the tube over extended periods of time.

When an accelelerated electron beam strikes a metal target in the anode, X-rays are generated within the beam focus. The generation efficiency is less than one percent. Most of the energy is dissipated as thermal energy and elevates the temperature of the metal target. In the case of sealed-off X-ray tubes for normal use, the allowable load ranges from 0.1 to 2.0 kW with focus sizes of 0.1–10 mm². Such a load is obtained by the forced cooling of the focused target face with flowing water.

The load per unit area of the focus, called brilliance, can be enhanced by the reduction of the focus area. A so-called fine focus X-ray tube (*Ehrenberg* and *Spear* [2.1]) is built on the basis of this principle. A focus size of about 0.1 mm × 1.0 mm is most often utilized for X-ray diffraction applications. The brilliance in this case is normally 1.5 kW mm^{-2} with a Cu target. Finer foci are also used occasionally for specific research requirements.

There is another way of increasing the brilliance. It consists of placing the focus always on a fresh, cooled face. This concept has led to the development of various X-ray tubes that have movable anodes. An attempt to achieve increased brilliance by moving the anode was made by *Wood* as early as 1897, as quoted by *Breton* [2.2]. A variety of ideas seems to have materialized since then by way of trial manufacture. In the course of time, *Bouwers* [2.3] made public for the first time the perfection of an X-ray tube equipped with a rotary target. This X-ray tube employed the principle of the induction motor for the anode rotation. The tube is said to withstand a load of 30 kV, 200 mA for 1/20 s and is used for instantaneous photography. It does not permit continuous operation because it has no mechanism that forcibly cools the anode. But it may be regarded as a prototype for X-ray diagnosis. The X-ray tube made by *Stintzing* [2.4] may be cited as another type. *Stintzing* constructed a spherical anode target designed so that it can independently rotate around the horizontal axis and the vertical axis, respectively. It is said to withstand an enormous load as great as 100 kV, 500 mA for operation of 20 s.

The nonartificially cooled anode remains completely in the high vacuum chamber. This is of merit from the structural viewpoint, but this type anode cannot withstand continuous operation over long hours because of its lack of a cooling mechanism. X-ray diffraction procedures, however, make it imperative that the X-ray tube permit continuous operation; therefore, the X-ray tube must be force-cooled.

Thus, in the case of moving anode X-ray tubes, the important technical problems are a) initiation of movement of the anode into the vacuum and b) a mechanism for forced cooling. Varied approaches have been made to resolve these problems.

For the exit location of the moving shaft through the vacuum wall, two methods of vacuum sealing have been tried:

1) A method having no sliding motion at the vacuum seal position.

This method provides the anode with mobility by connecting it to the tube wall by means of flexible bellows. This technique, by which the anode oscillates, precesses, or reciprocates on the plane, was pioneered by early workers in the field. *Dumond* and *Youtz* [2.5] employed the precessional movement for the target face of the anode and built an X-ray tube which could withstand a continuous load of 30 kW at 287 kV and 105 mA. The brilliance of the load was 0.38 kW mm^{-2}. An X-ray tube of a similar system but more compactly designed was made by *Hosemann* [2.6], in which the target was aluminum and the brilliance was 0.082 kW mm^{-2}, a value that is rather small.

Then, *Edamoto* [2.7] followed. In addition to the precessional motion of the anode, he incorporated a motion of the focus that generated the Archimedian spiral locus on the target face. The tube was operated for long hours with the focus of 0.1–0.3 mm in diameter and a load of 30 kV and 10–20 mA (Cu target) without any appreciable damage on the target face.

Astbury and *Mac Authur* [2.8] constructed another X-ray tube, in which the target was made of a pipe of rectangular cross section. By applying flexible bellows to both ends, they produced a vacuum seal and made the anode movable. They obtained a brilliance of 0.27 kW mm^{-2} with a load of 28 kV, 44 mA.

2) The method of applying vacuum sealing to the anode rotary shaft.

The easiest way of raising the relative velocity between the focus and the target face is to use a disk or cylindrical target and have the target rotated around its axis. In this method, it is necessary to apply vacuum sealing to the rotary shaft at the inlet to the vacuum chamber. A stream of cold water must be circulated through the anode for cooling.

X-ray tubes in which the vacuum seal is made by the mechanical fitting of the rotary shaft on the bearing were made by *Stintzing* [2.9], *Müller* [2.10], and *Fournier* and *Mathieu* [2.11], but these do not seem to have performed satisfactorily. *Clay* [2.12], collaborating with *Müller*, made the vacuum seal to the rotary shaft of the anode by means of three sets of Vaseline-applied leather gaskets. The target, which had a 5.8 cm effective diameter, was rotated at the rate of 2000 rpm with a 1-HP motor. A load of 30 kV, 150 mA was available with a Cu target, and 1000 hours operation in two years is reported. Then, *Clay* and *Müller* [2.13] made further improvements and built an X-ray tube equipped with an Fe target 20 inches in diameter, which permits rotation at 2000 rpm with a 5 HP DC motor. They obtained an allowable load of 2 A at 25 kV and 1 A at 50 kV with the focus of 1–2 mm × 30 mm.

A method of sealing the rotary shaft with mercury is reported by *Astbury* and *Preston* [2.14], and *Nishiyama* [2.15] independently, six years apart. The method features frictionless sealing of the rotary shaft.

Linnitzki and *Gorski* [2.16] applied the principle of the molecular pump to the area between the rotary part of the anode and the tube wall to make it airtight. With this tube, the anode rotation also acts as a vacuum pump and therefore there is no need for a diffusion pump to obtain the high vacuum. Moreover, the seal portion of the rotary anode offers the possiblity of being a no-contact type. However, to have the anode act as a molecular pump, exceedingly high speed rotation (> 10000 rpm) and accordingly a high degree of precision machining are required.

The so-called *Wilson* seal [2.17] was developed in recent years for vacuum sealing the rotary shaft. *Miyake* and *Hoshino* [2.18], *Yoneda* et al. [2.19], and *Kiyono* et al. [2.20] experimentally made X-ray tubes by using this sealing technique. According to *Miyake*, the anode rotary shaft 23 mm in diameter was sealed, and the seal life proved to be as long as 300 h at 1200 rpm. With the advent of this sealing technique, the construction of the sealed portion has become simple and permits easy handling. At the same time, the seal performance has been upgraded.

With an oil seal, *Taylor* [2.21] constructed a compact rotating anode which can be easily assembled and disassembled, unlike previous tubes. A brilliance of 1 kW mm^{-2} with an allowable load of 5 kW was obtained. *Davies* [2.22] further reported another X-ray tube of the same sort, in which the treatment of scattered electrons from the target is included. *Y. Shimura* commercialized a rotary anode X-ray generator provided with a rotary anode of the same quality as *Taylor*'s.

With the rotary anode tube, the larger the focus, the better the efficiency is obtainable as compared with X-ray generators having a stationary-type target. However, *Gay* et al. [2.23] combined a fine focus of 0.15 mm × 1.5 mm with a rotary anode, and thereby improved the brilliance to 4.4 kW mm^{-2}.

Aiming at a high brilliance and high power X-ray source, there are even some attempts to make use of synchrotron radiation at present. To this end, however, the only tool that permits easy use in the X-ray diffraction application and is available to researchers may be the rotary anode X-ray generator. Among varied types of this X-ray tube, it is very likely that *Taylor*'s type applied with further technical improvement will become most popular. X-ray generators that yield far greater output or super-high brilliance are also on the market, but technical improvements are still being made to enhance their performance.

In the following sections, theories and techniques for obtaining high brilliance or high power in X-ray tubes and the situation of X-ray generators in recent years will be surveyed. In Section 2.2, theories of temperature rise in foci due to electron beam bombardment for various types of targets are briefly described. The thermal strain induced in foci is emphasized as a factor restricting the increment of load available. Also, X-ray intensities emitted from targets are discussed in connection with the emitting directions. In Section 2.3, some of the techniques to be noted are referred to for manufacturing X-ray tubes of high brilliance or high power. In Sections 2.4 and 2.5, high brilliance or high power X-ray generators which are used at present and are manufactured for a special purpose are described.

2.2 Brilliance of X-Ray Sources on Stationary and Moving Targets

2.2.1 Brilliance of Foci for Stationary Targets

The greatest factor for determining an allowable load on a stationary target is that of the temperature of the target which is elevated by the bombardment of electrons with ensuing melting. *Müller* [2.24] calculated the temperature rise of the target, which is assumed to be a disk as shown in Fig. 2.1, and obtained the allowable load. With respect to the circular focus, the maximum input power is

$$W_{stat} = 15.8 \times \frac{(T_m - T_0)\delta k}{\theta}$$

$$\theta = \int_0^\infty \exp(-x^2)\tanh(\beta x)dx, \quad \beta = 2\sqrt{\ln 2}\frac{d}{\delta} \tag{2.1}$$

where T_m is the melting point in °C, k is the thermal conductivity in cal s^{-1} cm^{-1} °C^{-1}, δ is the radius of focus spot in cm, and a and d are the radius and the thickness of the target in cm, respectively. The distribution of electron beams

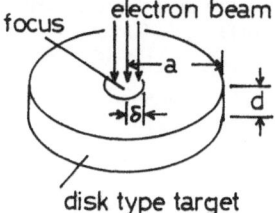

focus   electron beam

disk type target

Fig. 2.1. Model for calculation of a temperature rise of a stationary target (after *Müller* [2.24])

Table 2.1. Relations between d/δ, β and θ defined in (2.1)

d/δ	$\beta = 2\sqrt{\ln 2}\, d/\delta$	$\theta = \int\limits_0^\infty \exp(-x^2)\tanh(\beta x)\, dx$
0	0	0
0.06	0.10	0.0498
0.3	0.50	0.235
1.5	2.50	0.632
3.0	5.00	0.751
6.0	10.0	0.817
∞	∞	0.886

at the focus is assumed to be Gaussian. Moreover, a is assumed to be very large compared with δ. Table 2.1 shows the relations between θ, β, and d/δ. Also, the relations between the focal size and the allowable load in the case of Cu targets, calculated from (2.1), are shown in Table 2.2.

These relations show that the maximum input power (allowable load) is approximately proportional to the focus size, but the brilliance is approximately inversely proportional to the focus size. Accordingly, the reduction of focus size is effective for increasing the brilliance.

The allowable load to elliptic foci (which is often the case with actual X-ray tubes) is given by (2.2) (*Müller* [2.25]).

$$W_{\text{stat}} = 17.8 \times k \times (T_{\text{m}} - T_0) \times \mu(\delta_1, \delta_2) \tag{2.2}$$

where $\mu(\delta_1, \delta_2)$ is the function of the longer radius and shorter radius of focus, and

$$a_1 = (\delta_1 + \delta_2)/2 \qquad b_1 = \sqrt{\delta_1 \delta_2}$$
$$a_2 = (a_1 + b_1)/2 \qquad b_2 = \sqrt{a_1 b_1}$$
$$\vdots \qquad\qquad\qquad \vdots$$
$$a_n = (a_{n-1} + b_{n-1})/2 \qquad b_n = \sqrt{a_{n-1} b_{n-1}}$$

$$\lim_{n=\infty} a_n = \lim_{n=\infty} b_n = \mu(\delta_1 \delta_2),$$

Table 2.2. Relations between the maximum input power and focus size in the case of circular foci on the stationary Cu target. $k = 0.79$ cal s^{-1} cm^{-1} °C, $T_m - T_0 = 1000$ °C, $d = 0.1$ cm

Focus size: 2δ [mm]	Maximum input power: W_{stat} [W]	Brilliance [kW mm^{-2}]
0.01	7.04	89.6
0.1	72.6	9.24
1.0	892	1.14

Table 2.3. Allowable power in the case of elliptic foci on a Cu target (constants used in the calculation are the same as those in the case of Table 2.2.)

Focus size $2\delta_1 \times 2\delta_2$ [mm^2]	Maximum input power W_{stat} [W]	Brilliance [kW mm^{-2}]
0.01 × 0.1	29.9	38.0
0.1 × 1	299	3.80
1 × 10	2990	0.380

in which case, $\delta_1, \delta_2 \ll d, \delta_1, \delta_2 \ll a$ are assumed to hold. The calculated value with the Cu target from (2.2) is shown in Table 2.3.

The calculated value of the allowable load of the focus of 100 μm × 1000μm^2 in size is 300 W. However, the load which does not cause roughness on the surface of the target made of anoxic copper is about 200 W by measurement, and this value is about 67% of the calculated value. Similar ratios are found for other focus sizes.

A foil target may also be used for small foci. In this case, X-rays are extracted from the electron incidence side of the foil as well as from the opposite side. Since divergent X-rays in a broad angle range are obtainable, these tubes are used for photography of pseudo-Kossel patterns and related phenomena. With the model illustrated in Fig. 2.2, consider that the circumference of the foil of radius r and thickness d is cooled to a temperature T_0 and that the heat is generated uniformly inside the cylinder of radius δ. The allowable load is then expressed by the following relation:

$$W_{stat} = 4\pi dk(T_m - T_0) / \left(1 + 2\ln\frac{r}{\delta}\right). \tag{2.3}$$

An example calculated by (2.3) is shown in Table 2.4. Equation (2.3) is derived on the assumption that the uniform heat generation takes place in the direction of the thickness of the foil. Actually, the electrons lose their energy within a few microns from the foil surface. Accordingly, computer calculations were made by assuming that the heat is generated at a depth of 3 μm from the surface, and that

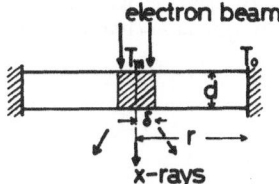

Fig. 2.2. Model for calculation of a temperature rise of a foil target

Table 2.4. Allowable power on the copper foil target, in the case of circular foci

Focus size, 2δ [μm]	Maximum input power W_{stat} [W]	Brilliance [kW mm^{-2}]
10	6.70 (5.2)[a]	85.3
100	10.7	1.36

[a]   As (2.3) is not fairly correct when $2\delta \leq d$, W_{stat} in the case of 10 μm focus was numerically computed more exactly and shown in the parentheses. Foil thickness is 20 μm.

the focus attains a Gaussian heat distribution. The result shows that the allowable load is 5.2 W with a 10 μm diameter focus. When actually fed with a load of 5.2 W, the target foil was pitted in about 15 min. In order to maintain the foil life for 48 hours, the load should be lowered to 3 W. Oxidation from the rear surface of the foil was first considered, as a possible cause for this pitting, but this was considered groundless because it was found that the same phenomenon occurs with a platinum foil target.

According to (2.3), the allowable load is proportional to the foil thickness. On the other hand, since X-rays are absorbed by the foil itself while passing through the foil, an appropriate thickness of the foil target should be $1/\mu$, where μ is the linear absorption coefficient of X-rays. As shown in Table 2.4, even if the focus is increased in size, there results merely a slight increase of the allowable load. Thus, the target is effective when the focus is less than several tens of microns.

When the focus becomes less than 1 μm, a cathode having a general thermal tungsten filament can no longer reach the allowable load for the target in terms of thermal electron feed capacity (*Cosslett* and *Nixon* [2.26]). To cope with the situation, a point cathode (*Dolan* and *Dyke* [2.27]) and a cathode employing LaB$_6$ (*Broers* [2.28]) have been developed, but these will not be discussed here as they deviate from the category of high brilliance X-ray sources.

2.2.2 Brilliance of Foci for Rotary Targets

Müller [2.24, 25, 29] calculated a formula of the allowable load in the case where a focus having a Gaussian exothermic distribution is assumed to move on a target of a semi-infinite dimension. Next came *Dumond* et al. [2.30] who

Table 2.5. Comparison between the temperature rises computed by *Ishimura*'s equation and (2.4) in the case of an Mo target. $2\delta = 0.05$ cm, $w = 600$ kW cm^{-2}, $D = 40$ cm, and $n = 1000$ rpm (after *Chikawa* [2.32])

Target thickness [cm]	*Ishimura*'s [°C]	(2.4) [°C]	Difference [%]
0.2	1704	1860	9.1
0.1	1691	1845	9.1
0.01	1686	1830	8.5

calculated the allowable load, assuming that a disk-shaped target with a finite thickness is rotated and is exposed repetitiously to the electron beam focus. In both cases, the conduction of heat is assumed to be perpendicular to the target surface. *Ishimura* et al. [2.31] considered that there should also be conduction of heat in the direction tangential to the target surface, and calculated the allowable load in detail for rotary targets.

Since the surface velocity of the rotating target is more than $10 \, \text{m s}^{-1}$, the elapsed time that a point on the target surface passes through the width of focus is as short as several tens of microseconds. The diffusion distance of heat for such a short time interval is very small compared with the width of focus. Therefore, the temperature on the target surface is considered to rise simultaneously over an extended region, and the thermal conduction in the direction tangential to the surface could be neglected. With this idea, *Chikawa* [2.32] substituted the integral for the solution in a form of infinite series, derived by *Ishimura* et al., and obtained the following equation.

$$T_{\text{m}} - T_0 = \frac{wd}{k} \frac{2\delta}{\pi D} + \frac{2w}{\pi} \sqrt{\frac{2\delta}{DnkC\varrho}} \tag{2.4}$$

where w is the energy of the electron beam per unit area and unit time, cal cm^{-2} s^{-1}, 2δ is the width of the electron beam in cm, D is the diameter of the target in cm, n is the number of revolution of the target, rounds s^{-1}, and C is the specific heat in cal g^{-1} °C^{-1}.

When $2\delta/D$ is small, the first term of (2.4) is negligible, and as a result, (2.4) will come to agree with *Müller*'s formula. *Oosterkamp* [2.33] also ignored conduction of heat in the direction tangential to the surface, and obtained (2.4) in an approximate way.

The first term corresponds to the temperature rise saved after cooling in a period of one revolution. The second term corresponds to the temperature rise while a point on the target surface passes through the electron beam. For rotary targets usually used, the first term is within 10 % of the second term. Accordingly, the calculation with only the second term could suffice for estimating an allowable load of a rotary target. *Chikawa* actually used both the equation by *Ishimura* et al., and (2.4), and obtained Table 2.5. His work shows the difference

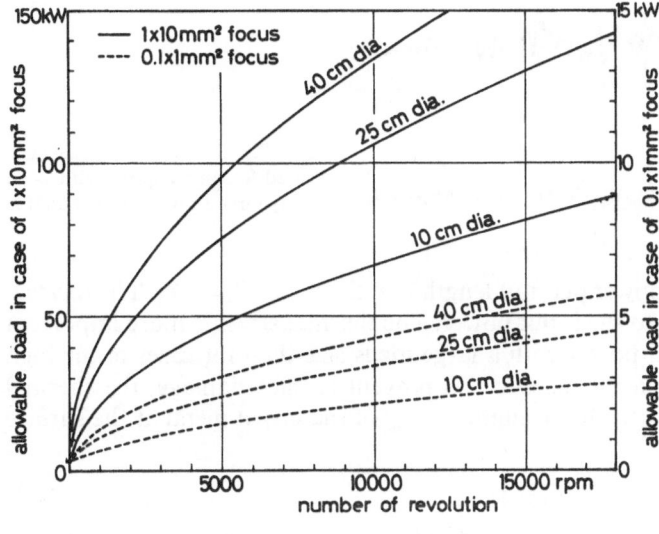

Fig. 2.3. Allowable load for rotating copper targets of various diameters calculated using (2.4)

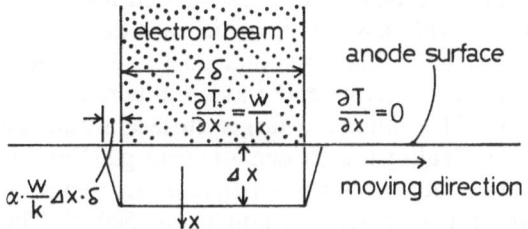

Fig. 2.4. Thermal strain in a surface of a rotary target (after *Chikawa* and *Fujimoto* [2.35])

between the two approaches to be within 10 %, which demonstrates that the flow of heat in the tangential direction to the target surface can indeed be neglected in practice.

Figure 2.3 shows an example of calculation of the allowable load with (2.4).

Thermal stress is another important factor which determines the allowable load for the target. *Sato* [2.34] calculated the thermal stress in a target and indicated that it is much higher than the actual material strength. *Sato*'s formula, however, is fairly difficult to understand. *Chikawa* and *Fujimoto* [2.35] approached this problem in the following way: The maximum temperature gradient on the surface is given by

$$-k\left(\frac{\partial T}{\partial x}\right)_{x=0} = \begin{cases} w & \text{inside of the electron beam} \\ 0 & \text{outside of the electron beam.} \end{cases} \tag{2.5}$$

Since $(\partial T/\partial x)_{x=0}=0$ outside of the electron beam, the shear strain $\alpha(w/k)\delta$ is induced at the edge of the part irradiated with the electron beam by the thermal expansion, as shown schematically in Fig. 2.4, where α is the coefficient of linear expansion. The shear stress σ is given with the shear modulus τ by

$$\sigma = \tau\alpha w\delta/k = (\tau\alpha/2k)W, \tag{2.6}$$

Cu

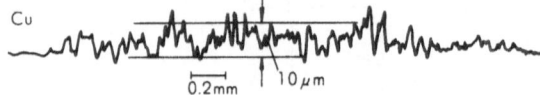

0.2mm   10 μm

Cu-Cr   (Cr : 1%)

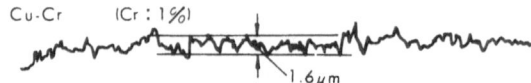

1.6 μm

Fig. 2.5. Surface roughness measured for pure copper and Cu–Cr alloy targets after 5 h of operation

where W is the input power per unit length into the focus, $W = 2\delta w$. The thermal stress is proportional to the input power. The thermal stress is more important for cases of large total power with a large focus size than for cases of the high brilliance with a smaller focus size. To prevent surface damage, the thermal stress should be less than the strength $\sigma_m(T_m)$ of the target metal at the surface temperature T_m,

$$\sigma \leqq \sigma_m. \tag{2.7}$$

As an example, *Chikawa* considered a Cu target. Using $D = 40$ cm, $d = 0.2$ cm, $n = 17$ rounds s^{-1}, $2\delta = 0.5$ mm and $w = 600$ kW cm^{-2}, (which are dimensions adopted for RU – 500 to be mentioned in Sec. 2.5.2), he obtained $T_m = 900\,°C$ and $\sigma = 50$ tons cm^{-2}. The strength of copper is 5 tons cm^{-2} at room temperature, and it decreases rapidly at 200 °C. Therefore, the target surface would be damaged. Instead of pure copper, Cu–Cr alloy was chosen for the target. For this alloy, the thermal conductivity is about 80 % of pure copper, and the strength is 7 tons cm^{-2} at room temperature and is nearly constant up to 500 °C. The surface damage and X-ray output intensities for the alloy and pure copper targets were compared after 5 hours of operation. Figure 2.5 shows the surface roughnesses for pure copper and for Cu–Cr alloy targets, which were measured to be about 10μm and 1.6μm, respectively (1.5μm for polished surfaces of both the targets before use). The X-ray output intensities of the alloy target were measured to be about 30 % and 15 % higher for the take-off angles of 3.3° and 5.3°, respectively, than those of pure copper targets. From these results, the long-term operation is expected to cause a much more drastic output decrease for the targets consisting of pure copper.

For Mo targets, the surface temperature and thermal stress were calculated to be $T_m = 1860\,°C$ and $\sigma_m = 60$ tons cm^{-2} with the same condition as mentioned above, which is comparable with a strength of 50 tons cm^{-2} at room temperature. No surface damage was found after operating for many hours.

2.2.3 X-Ray Yield

Green [2.36] examined the question of how many X-rays will be generated when one electron impinges on the target. When no absorption by the target itself is assumed, the number of characteristic X-rays, N, is given by

$$N = K_0(E_0 - E_x)^{1.63}, \tag{2.8}$$

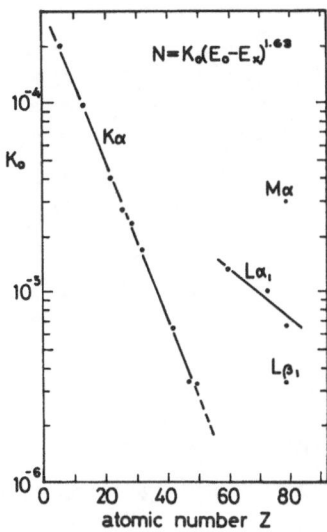

Fig. 2.6. Quantum efficiencies K_0 of X-ray production for various transitions as a function of atomic number (after *Green* [2.36])

where E_0 is the energy of incident electrons in kV, E_x is the minimum exciting voltage for characteristic X-rays in kV, and K_0 is the proportional constant as shown in Fig. 2.6.

When electrons are actually projected on a target, however, some of them infiltrate into the target and generate X-rays. Consequently, X-rays emitted from these electrons would be absorbed by the target material itself until they emerge out of the target surface. The amount of absorption depends on the target material, the incident energy of electrons, and the angle γ between the target surface and the direction of X-rays emitted. Let the number of X-rays per unit solid angle, generated from the target, be given as $N/4\pi$ and the number of X-rays per unit solid angle, emitted in the direction of the angle γ, be given as $N(\gamma)/4\pi$. Then, these must be related by

$$N(\gamma)/4\pi = f(\gamma)N/4\pi , \tag{2.9}$$

where $f(\gamma)$ represents the correction of absorption by the target itself. Figure 2.7 shows values of $f(\gamma)$ for Al K_α, Cu K_α, and Mo K_α measured by *Green*.

2.3 Technical Problems

2.3.1 Electron Guns

An electron gun designed to provide a sufficient flow of emitting thermal electrons according to focus sizes is one of the technical problems to be addressed in the design of X-ray generators. Fortunately, so far as electron guns

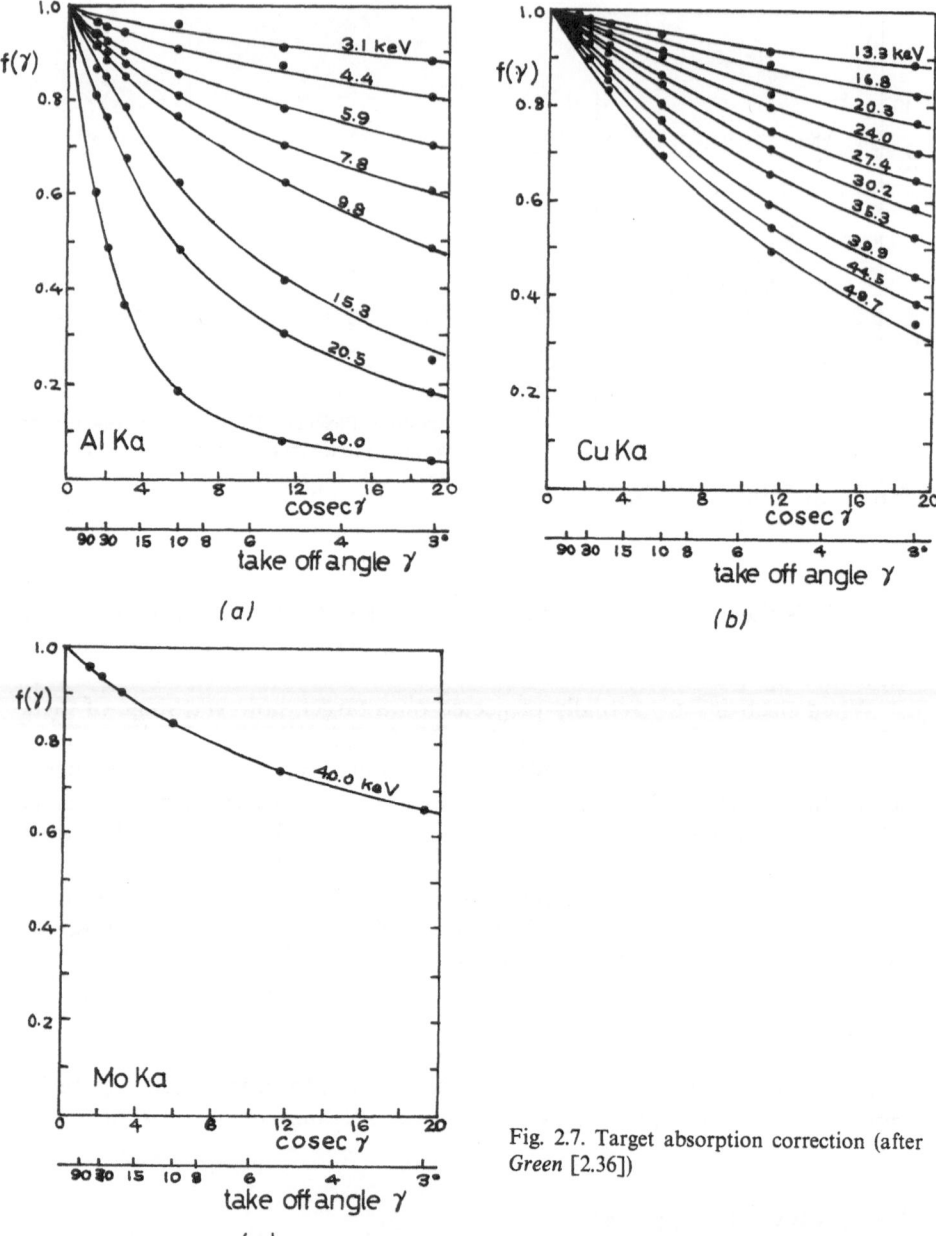

Fig. 2.7. Target absorption correction (after *Green* [2.36])

for a microfocus are concerned, axially symmetrical highly brilliant electron guns have already been developed in the field of electron microscopy, and this technique is applicable to the problem. However, there is an essentially different point with respect to the use for X-rays. While the performance of electron guns required for the electron microscope is to provide electron beams of sufficiently

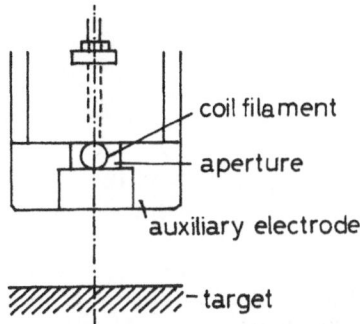

coil filament

aperture

auxiliary electrode

target

Fig. 2.8. Cross section of an electric gun used in a high brilliance X-ray tube

high density per unit area and unit solid angle at the crossover, the requisite for electron guns for X-ray tubes is only high electron density per unit area within the focus on the target. Hence, the opening angle of electron beams with electron guns for X-rays is generally broader than for that of the electron microscope.

Moreover, since the time for photography with the electron microscope is relatively short, that is, in the order of seconds, the filament is designed with its life aimed at about 50 hours at the heating temperature of about 2700 K, whereas, in the case of X-ray tubes, it is necessary to consider 500–1000 h operation with a filament temperature of about 2600 K.

The size of microfocus with X-ray tubes suitable for practical use is approximately 0.1 mm × 1.0 mm. On special occasions, finer foci are used. Generally speaking, however, it is practically impossible to construct or operate X-ray cameras which could utilize a focus smaller than this value (*Guinier* [2.37]). As an electron gun for a microfocus, the gun based on the studies by *Ehrenberg* and *Spear* [2.1] is often used. Although this gun has a drawback in that the current value is only about 1 mA, it can produce a 40 μm × 200 μm focus.

In the case of a microfocus with a rotary anode, increased load is feasible, as already mentioned in Section 2.1. Efficient guns must thus be developed. In order to obtain a sufficient current (about 20 mA to 60 mA), the effective area of the line filament is too small for emitting required thermal electrons. A coil filament is accordingly used to broaden such an area. By using the electron gun illustrated in Fig. 2.8, although it seems rather primitive, the aimed focus size and the current can be obtained by restricting the surface area of coil filament emitting electrons by the size of aperture around the coil filament, and by using a modified well-like, auxiliary electrode *Kuntke* [2.38] applied the bias voltage for focusing the electron beam. Today, electron guns for 1.2 kW (12 kW mm^{-2}) and 3.5 kW (35 kW mm^{-2}) with a 0.1 mm × 1.0 mm focus have been developed by the present authors' group. These electron guns are designed in the space charge restricted region and therefore are usually applicable with a tube voltage from 45 to 60 kV.

Electron guns having obtainable current from 200 to 1000 mA with a 0.5 mm × 10 mm focus or 1 mm × 10 mm focus required efforts towards increased emission of effective thermal electrons under the design concept of the

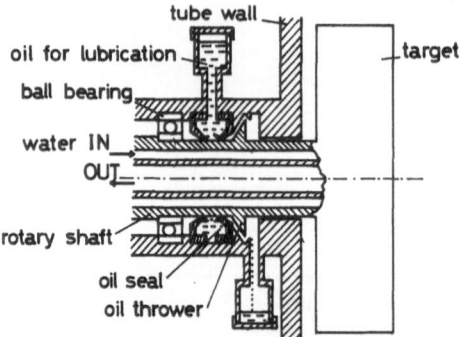

tube wall

oil for lubrication

ball bearing

water IN

OUT

rotary shaft

oil seal

oil thrower

target

Fig. 2.9. Principal diagram of rotary anode targets

aforementioned electron guns. These concepts have been put to practical use at present (*Shimura* et al. [2.39]).

Theoretical design of an electron gun that can yield a practicable electron flow is a task of extreme difficulty. The present situation is such that the gun shape is determined only through experience and trial-and-error experimentation.

With voltages as low as 10–20 kV as in the case of X-ray tubes designed to generate soft X-rays, the electron gun will instantly approach the cut-off state at a small current due to the space charge. Thus a large current is not achievable. To cope with this, a Pierce-type electron gun [2.40] designed for low voltage and large current is used. The present authors and their co-worker have obtained 1 A for soft X-ray generators with this type of electron gun.

2.3.2 Construction of Rotary Targets

A rotary anode which has a cylindrical face target is now in practical use, as mentioned in Section 2.1. Figure 2.9 shows the principal diagram of this configuration. The essential elements of this design are the cylindrical rotary target, a water circulating mechanism for cooling the target, and a vacuum sealing mechanism. Since Cu K_α characteristic X-rays find the highest utility in X-ray diffraction and in view of the high thermal conductivity of copper, it is customary that the target drum be made of vacuum-dissolved anoxic copper. To produce a Co or Cr target, cobalt or chromium should be plated on the Cu cylinder surface to a thickness of 20 to 100 µm. On the other hand, to produce a Mo target, molybdenum itself is used to fabricate the entire cylinder. Since Mo is apt to dissolve in water, the face exposed to a flow of water is covered with Ni wax.

Aluminum is often used as a target for soft X-rays. For the use, Al is deposited on the Cu cylinder surface to a thickness of 20–100 µm by the method of ion plating.

The allowable load to the rotating anode target is determined by the melting point of the cylinder surface metal to be exposed to the electron beam, by the

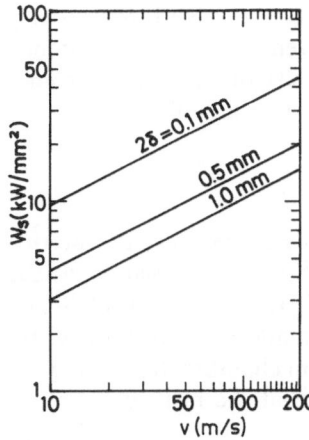

Fig. 2.10. Relation between the circumferential velocities of target and brilliancies in case of a Cu target

vapor pressure, and by the extent of thermal-fatigue breakdown. For instance, when a pure Cu target of 10 cm diameter is assumed to be rotated at 2500 rpm with a 0.5 mm × 10 mm focus, the target should be usable with a load up to 25 kW without melting. In a practical case, however, 50% of the allowable load to be determined by the melting point is set as an empirically allowable load so as to keep the target surface from being roughened by the thermal fatigue breakdown of the surface layer. The influence on X-ray intensities due to the thermal fatigue was discussed in Section 2.2.2.

When the allowable load is required to be increased further with the circumferential velocity remaining constant, a metal that exhibits high-tension resistance at high temperatures should be employed. An alloy fulfills the conditions for such a material as mentioned in Section 2.2.2. In the selection, however, it is necessary that the intensity of X-ray spectra of the added elements should be negligible compared with those of the base metal.

As a favorable pure metal for targets, molybdenum may be considered. We have already mentioned that Mo has a drawback of easy dissolution to water but has the virtue of low vapor pressure at high temperatures and no appreciable decrease of tension resistance even at high temperatures. This metal, therefore, permits use with a load up to the allowable limit of the melting point.

There is a problem for consideration as to how much the allowable load can be raised in relation to the machine structure of the rotary target. As mentioned in Section 2.2, we may safely say that the allowable load limited by the melting point should be after all determined by the velocity of the target surface that passes through the electron beam focus. Figure 2.10 shows such allowable loads in relation to the brilliance and the circumferential velocity of the target. Needless to say, the circumference velocity is determined by the target diameter and the revolution frequency. It is not so easy, however, to increase this circumferential velocity as freely as desired, due to such factors as the mechanism for feeding sufficient cooling water to the interior of the target and the mechanical strength of metals and their processing, assembling technique, as well as economical costs. These factors should be examined respectively according to the purpose of use. At present, when disregarding economy, it may be possible to make available a circumferential velocity of 200 m s^{-1}. From a realistic point of view, 130 m s^{-1} to 150 m s^{-1} may be the limit.

2.3.3 Vacuum Seals

The vacuum sealing for the rotary shaft will next be discussed. Since the interior of the target has to be cooled with flowing water, this will necessitate, at the current stage of technique, leading the rotary shaft through the wall of the vacuum chamber out into air. The vacuum sealing is made at the location where the rotary shaft passes through the wall of the vacuum chamber. Reviewing the vacuum sealing, it may be no exaggeration to say that the history of rotary anodes is that of the sealing technique.

As methods of vacuum sealing technically available at present, there are the mechanical seal, magnetic fluid seal, and oil seal as representative ones. Among them, in the case of the mechanical seal, the faster the speed at the sealed portion, the larger the cooling mechanism required for the sealed portion, and therefore a special-purpose cooling-oil circulation system will be needed. Moreover, as the starting and stopping of operation are repeated, the seal surface will wear out rapidly, and seal replacement is required.

The magnetic seal has a characteristic of generating heat for itself due to the viscosity of the seal material, and the viscosity will decrease drastically as the revolutions increase. Thus, there is a limit on the circumferential velocity at the sealed portion. The limit seems to be less than 7 m s^{-1}. This seal may permit use up to about 5000 rpm with a diameter of 25 mm. The magnetic seal is, however, not applicable at present for vacuum sealing of the rotary shaft of high-power or high-brilliance X-ray targets which necessitates a velocity of about 10000 rpm.

The oil seal has the widest applicability. The authors' experience indicates that it can practically serve as a vacuum seal with a circumferential velocity of more than 10 m s^{-1}. The maintenance cost is also very low for this type of seal. In addition, the use of oil seals makes it easy for users themselves to replace them. The authors and their co-workers have succeeded in obtaining a prolonged life of 2000 h for an oil vacuum seal at 10000 rpm with a rotary shaft of 35 mm in diameter.

2.3.4 Cooling Water

The rate of flowing water to cool the target is determined by setting the average temperature rise of the flowing water at the exit of the target at 10–15 °C. The flow rate is about 6 l min^{-1} with a 12-kW X-ray generator and about 20 l min^{-1} with a 30-kW generator. In the case of the latter flow rate, a cooling circulation system will be required to keep water from being thrown away and wasted. The quality of the water should be noted. Distilled water or soft water of less than 1 ppm chlorine ions is recommended to avoid corroding the target metals.

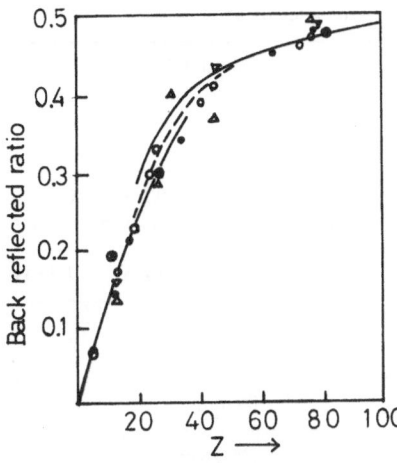

Fig. 2.11. Ratios of back-scattering electrons (after *Archard* [2.41])

2.3.5 Influences of Back-Scattered Electrons

Lastly, the action of recoil electrons from the target surface will be discussed. Not all electron beams from the cathode are absorbed by the target when they reach it. Some of them are ejected into the vacuum from the target surface. Figure 2.11 shows that the ratio of the number of electrons scattered into the vacuum to that of electrons incident upon the target depends on atomic numbers (*Archard* [2.41]). This ratio is independent of the voltage applied to the tube, which is the reason why Mo targets can be used with the allowable load determined by calculation using the temperature of the melting point, as mentioned before. The energy of electrons scattered from the target into the vacuum ranges from the energy equivalent to the applied voltage to that of secondary electrons. Consequently, these electrons heat the X-ray tube window and the tube wall. Especially, a part of the wall near the focus and the window suffers from intense heat. Moreover, in the case where a vacuum system having an oil diffusion pump is used, the back-flow oil vapor might be decomposed by heating due to the scattered electrons, depending on the performance of the diffusion pump. This will not only result in contaminating the target surface and the tube wall but will also make the vacuum unstable and cause electric discharge inside the tube. Accordingly, it is necessary to occasionally clean the target surface and the wall inside the tube.

2.4 Contemporary Rotary Anode and Microfocus X-Ray Generators

2.4.1 Rotary Anode X-Ray Generators

Rotary anode X-ray generators that are most widely used now may be *Rigaku*'s Model RU-200 and *Elliott*'s Model GX-6. Table 2.6 illustrates the characteristics of both models. RU-200 provides the foci ranging the minimum of 0.1 mm

Table 2.6. Specifications of contemporary rotary anode X-ray generators

| | Rigaku, RU–200 | | | | Elliott, GX–6 | | |
	Normal type	High brilliance type					
Focus size	0.5 mm × 10 mm	0.3 mm × 3 mm	0.2 mm × 2 mm	0.1 mm × 1 mm	0.3 mm × 3 mm	0.2 mm × 2 mm	0.1 mm × 1 mm
Max. load	12 kW	5.4 kW	3 kW	1.2 kW	4 kW	3 kW	1.2 kW
Max. tube voltage and current	60 kV, 200 mA				50 kV, 80 mA		
Target dia.	99 mm				89 mm		
Revolution of target	2500 rpm	6000 rpm			6000 rpm		
Vacuum seal	Oil seal				Oil seal		

× 1 mm to the maximum of 0.5 mm × 10 mm. The foci of less than 0.3 mm × 3 mm are suitable for X-ray diffraction instruments such as the precession camera, *Franks'* camera, the point focusing camera, the *Lang* camera, and the microdiffractometer of *Rigaku*, as well as the microbeam camera, and other applications. On the other hand, the 0.5 mm × 10 mm focus is designed to fit to usual X-ray cameras and diffractometers. The available output of 12 kW in this case means that the X-ray intensity is 6–8 times higher than that of the sealed-off X-ray tube having comparable focal size.

Figure 2.12 shows the cross section of the RU-200 target. The target is cooled with a stream of water of about 6 l min^{-1}. An oil seal is used for the vacuum sealing. The sealed portion is conveniently arranged so as to permit easy installation, removal, and replacement. The target is 99 mm in diameter and is light weight and easy to handle as compared with the high-power rotary target to be referred to later. Focus size is changeable by replacing the cathode (the tungsten coil filament and focusing electrode) en bloc, and therefore the adjustment work for filament position is unnecessary. During operation, the X-ray tube is always evacuated by means of the oil diffusion pump and oil rotary pump. The valve operation of the vacuum system is automated.

2.4.2 Microfocus X-Ray Generators

As mentioned in the preceding section, the foci down to 0.1 mm × 1.0 mm are made on the rotary target. As for further smaller foci, X-ray generators designed specifically for the purpose are available. One type is such that the electron beam is converged by the operation of a three-electrode electron gun. Another type uses an electromagnetic lens. In the former, the electron beam may be stopped

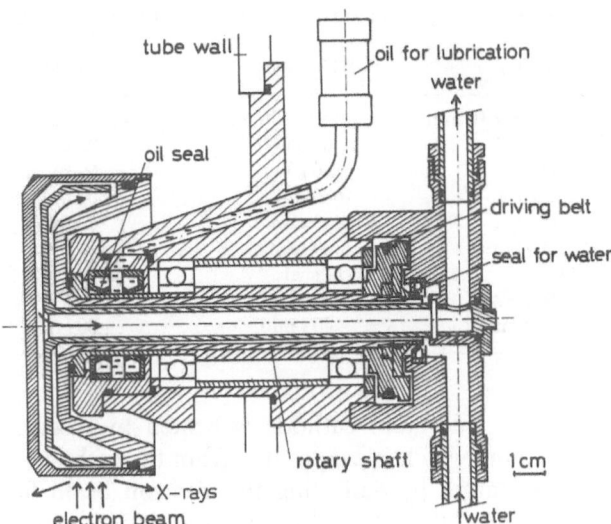

tube wall

oil for lubrication

water

oil seal

driving belt

seal for water

rotary shaft

1 cm

X-rays

electron beam

water

Fig. 2.12. Cross section of the target of Model RU-200

down to about 30 µm in diameter. In the latter, the electron beam may be easily stopped down to about 5 µm in diameter. These generators have been put on the market by Hilger, Rigaku, and others.

To obtain an X-ray source of below 1 µm, a two-stage electromagnetic lens is usually employed (*Cosslett* et al. [2.26]), but such a device has not yet come into general use commercially.

2.5 Special X-Ray Generators

2.5.1 Soft X-Ray Generators

Soft X-rays are mainly used for the excitation of photoelectrons in ESCA or for X-ray lithography. *Siegbahn* et al. [2.42] produced an X-ray source for Al K_α rays. A high-tensile aluminum alloy is used for the target, which is 11 cm in diameter and rotates at 5000 rpm. The load is 6 kW and the focus 4 mm × 2 mm. A mechanical seal is used for the vacuum sealing, and the target is driven by an air turbine motor. The electron gun is the Pierce type (*Pierce* [2.40]), and the cathode is made of tungsten and is heated by electron bombardment.

The present authors have also been involved in the manufacture of an X-ray generator for Al K_α rays. The target of 99 mm in diameter rotates at 6000 rpm. The focus is 5 mm × 10 mm and the load is 20 kW. To make the target, aluminum was ion-plated on a copper target cylinder to a thickness of about 20 µm. The target may be also made of pure aluminum. However, to increase the load, a high-tensile alloy is judged to be superior. Aluminum, whether it is pure or an alloy, suffers from the problem of corrosion by water. The electron gun is of the Pierce type, and the shape of the electrode was optimized with a computer. The cathode, which is made of a circular tantalum piece, emits an electron beam of

Table 2.7. Specification of super high-power X-ray generators, RU–500, RU–1000 and RU–1500

	RU–500	RU–1000	RU–1500
Focus size	0.5 mm × 10 mm	1 mm × 10 mm	1 mm × 10 mm
Max. load	30 kW	60 kW	90 kW
Max. tube voltage	60 kV	60 kV	60 kV
and current	500 mA	1000 mA	1 500 mA
Dia. of target	400 mm	400 mm	250 mm
Revolution	1250 rpm	2500 rpm	10 000 rpm
Vacuum seal	Oil seal	Oil seal	Oil seal
Total electric input power	50 kVA	93 kVA	150 kVA
Flow rate of cooling water	20 l min^{-1}	40 l min^{-1}	60 l min^{-1}

1 A from its 25 mm diameter. The tantalum cathode is heated by electron bombardment. Condensed tantalum, which is evaporated from the cathode, is prevented from adhering to the target by deflecting the electron beam 90° electrostatically. For vacuum sealing of the rotary shaft, an oil seal is employed. Furthermore, to keep the target surface from being contaminated by carbon, an oil diffusion pump of little back diffusion is carefully used, and the vapor from the sealing portion is prevented from entering the vacuum chamber by constructing an additional clearance seal between the chamber wall and the oil seal. The X-ray window has a Be foil 25 μm thick, which can withstand the atmosphere, with a diameter as large as up to 20 mm. The manufacture of another soft X-ray generator which should have a target of diameter 250 mm to rotate at 8000 rpm is presently under way.

2.5.2 Super High-Power X-Ray Generators

RU-200 and other similar X-ray generators have contributed greatly to the reduction of time required for X-ray diffraction of specimens under static conditions. X-ray intensities available from these generators, however, are still insufficient for recording under dynamic conditions or for observations by direct viewing by X-ray television. Rigaku's Model RU-500, RU-1000, and RU-1500 are designed to meet requirements such as dynamic observations (see Chap. 6 of this book). The major characteristics of these generators are illustrated in Table 2.7. To aim at higher allowable loads, the circumferential velocities of targets are increased by using targets of a large diameter or by increasing the number of revolution.

  Chikawa [2.43] conducted, for the first time, the direct observation of the motion of dislocations by the use of an RU-500 in combination with a video system. Figure 2.13 shows a video display of dislocation images obtained with this system; more details will be discussed by *Hartmann* in Chapter 6.

  RU-1500 is an apparatus under design process at present. The design concept is based on the extreme brilliance X-ray generator, FR, to be discussed in the following.

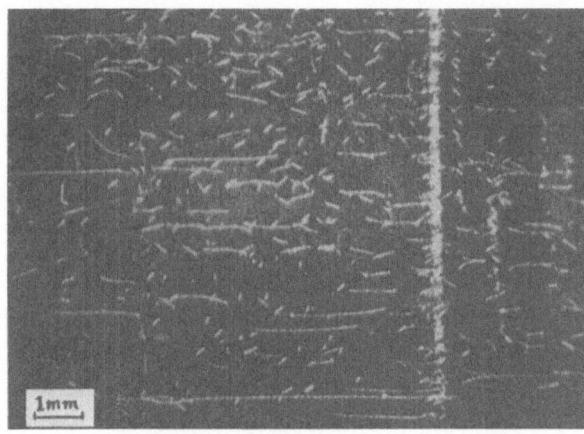

Fig. 2.13. Video images of dislocations in a silicon wafer obtained by RU-500 and Rigaku X-ray topograph imaging system. Mo-target, 60 kV,   500 mA.   (220) reflection

2.5.3 Extreme Brilliance X-Ray Generators

Organic materials such as muscles or polymers have, in general, wide lattice interplanar spacings, resulting in concentration of the associated X-ray diffraction spots into a narrow range of just a few degrees. Dispersion of these diffraction spots into a broader angular range would require the use of X-rays of longer wavelength, but this would cause the high absorption of X-rays in specimens, leading to difficulty in the specimen treatment. Improved resolution is achieved in a system in which a camera has an optical geometry to converge X-ray beams with a total reflection mirror and a curved crystal monochromator, using a microfocus X-ray source. For this purpose, the 0.1 mm × 1.0 mm focus of the aforementioned rotating anode X-ray tube such as RU-200 is used. However, organic specimens have generally weak diffraction, and moreover, the long exposure is impracticable due to their live condition. Also, from a standpoint of obtaining diffraction photographs under dynamic conditions, a further powerful X-ray generator is desired.

Figure 2.14 shows an external view and a cross section of the target of Rigaku FR (*Kozaki* et al. [2.44]). The target is 25 cm in diameter and rotates at 9000 rpm. Copper material of as little as 1 gram in mass on the surface of this target becomes a virtual weight corresponding to as much as 11.3 kg in the target rotation, because of the centrifugal force. In order to bear the resulting stress in each part of the target, a high tensile copper alloy is used as mentioned in Section 2.2.2. The resonance rotational frequency of the target is approximately 17 000 rpm. This is a result of thorough considerations so as to keep the resonance frequency far removed from the number of revolutions of the target. Consideration has also been given to the target diameter from the viewpoint that a large-sized target would entail a sharp rise in the production cost. The oil seals are used for vacuum sealing at the rotary shaft and their life is about 2000 hours. The focal size is 0.1 mm × 1.0 mm and the maximum load is 3.5 kW, and the

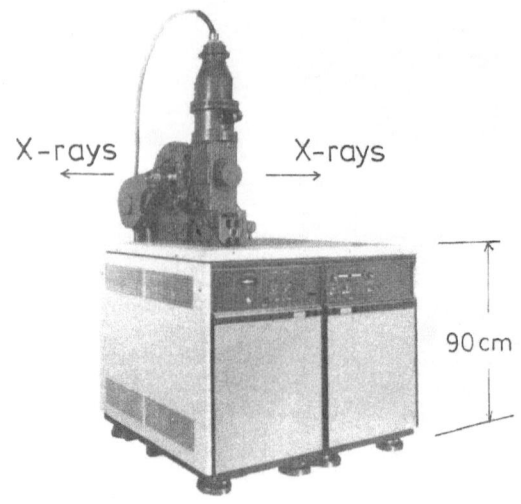

Fig. 2.14a. External view of the extreme brilliance X-ray generator, FR

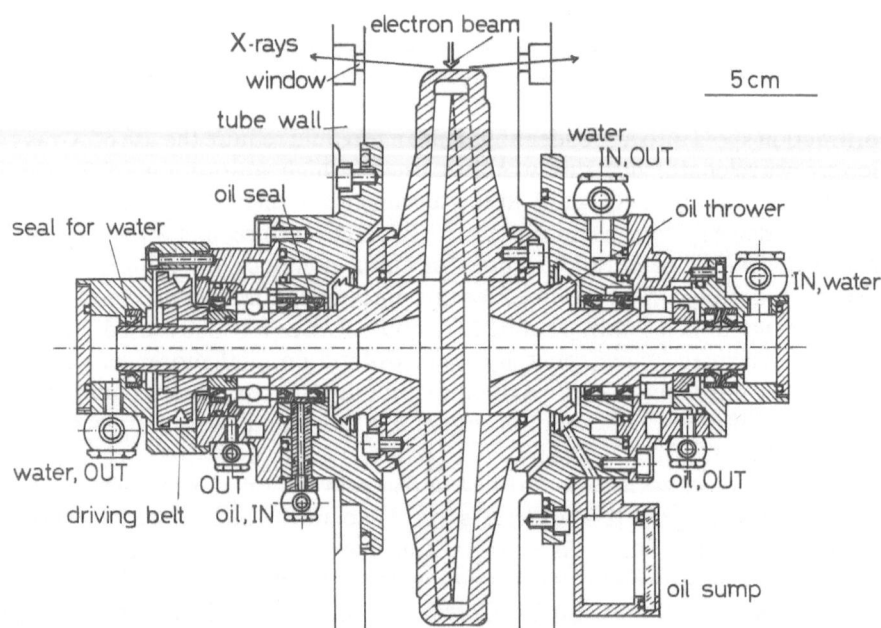

Fig. 2.14b. Cross section of the target of FR

brilliance is 35 kW mm^{-2}. X-rays are obtainable from two windows in the horizontal direction.

An example of a photograph taken with this generator is shown in Fig. 2.15. The specimen is a wet, raw tendon collagen taken from legs of fowl. Although the reproduced photograph is obtained by a 20-min exposure, the diffraction

20   50 100 400 400 100   50   20Å

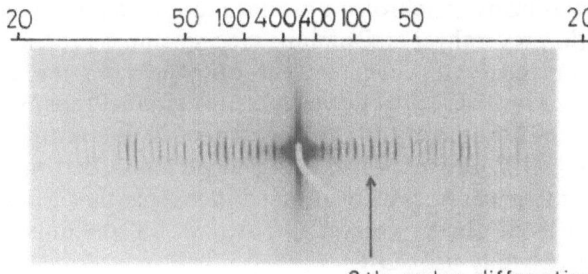

9th order diffraction

Fig. 2.15. Small angle diffraction photograph of native, wet fowl tendon collagen by the use of FR and point focusing camera. Cu-target, 20 min exposure

pattern up to 9th order can be obtained by only 0.5-min exposure. The camera is a Rigaku point focusing camera. In this camera the specimen-to-film distance is 300 mm, and the maximum measurable lattice constant is about 660 Å with Cu K_α radiations.

2.5.4 Pulsed X-Ray Generators

When extremely rapid changes like the phenomena of explosions, for instance, must be observed, the X-ray generators so far mentioned, which generate X-ray radiation in a continuous way, are no longer suitable. To fulfill such requirements, pulsed X-ray sources obtainable by putting to work as much power as possible in an extremely short time have been developed [2.45–48]. These devices made X-ray diffraction photographs which were taken under static conditions of the specimens.

Dantzig and Green [2.49] combined the pulsed X-ray source developed by Charbonnier with an image intensifier and obtained Laue photographs of aluminum and sapphire. Johnson et al. [2.50, 51] developed a pulsed X-ray generator of 50 kV, 50 mA, and 40 ns in pulse width, and also succeeded in synchronizating the generation of X-rays with that of a shock wave in a specimen. They thereby obtained diffraction photographs of the specimen where the plane of the shock wave is just traversing the specimen.

2.5.5 Outlook

In this chapter we have described X-ray generators having high performance. It seems that at present it is technically soundest to rotate an anode at high revolution velocity for obtaining X-ray tubes with high brilliance or high power. Since the allowable load of rotary targets is considered to be proportional to $\sqrt{D \cdot n}$ as mentioned in Sections 2.2.2 or 2.3.2, the circumference velocity of rotary targets must still be increased to obtain far superior performance. In Models RU-1500 or FR, the circumference velocity of the target has already reached 130 m s^{-1}. Accordingly, even if we desire to obtain a generator with a power 10 times greater than that of the RU-1500, it is impossible unless a circumference velocity of 130 m s$^{-1} \times 100$ m s^{-1} can be achieved.

Huge X-ray generators utilizing synchrotron orbital radiations (SOR) will come into practical use within several years. Opinions are sometimes presented that high-performance X-ray generators will become obsolete after the appearance of SOR generators. However, SOR generators are extremely costly to install and operate and it is clear that many research workers will not be able to use these generators as they wish. It is expected as a general trend that high-brilliance or high-power X-ray generators as mentioned in this chapter will be more exclusively for the use of small research groups or for preliminary experiments in order to successfully use SOR.

Acknowledgments. The authors are greatly indebted to Dr. *J. Chikawa* for his active cooperation in the early period of development of high-performance X-ray generators, and for his kind advice ever since. They also wish to acknowledge the authors from whose papers they have quoted data.

References

2.1   W.Ehrenberg, W.R.Spear: Proc. Phys. Soc. (London) B**64**, 67 (1951)
2.2   Breton: La revue scient. et industr. de lárnée (1898/99)
2.3   A.Bouwers: Physica **10**, 125 (1930)
2.4   H.Stintzing: Metallwirtschaft **17**, 761 (1938); **20**, 45 (1941)
2.5   J.W.M.DuMond, J.P.Youtz: Rev. Sci. Instr. **8**, 291 (1937)
2.6   R.Hosemann: Z. Tech. Phys. **20**, 203 (1939)
2.7   E.Edamoto: X-sen **3**, 5 (1942) (in Japanese)
2.8   W.T.Astbury, I. MacArthur: Nature **155**, 108 (1945)
2.9   H.Stintzing: Z. Phys. **27**, 844 (1926)
2.10 A.Müller: Proc. Roy. Soc. **125**, 507 (1929)
2.11 G.Fournier, J.Mathieu: J.Phys. **8**, 177 (1937)
2.12 R.E.Clay: Proc. Phys. Soc. (London) **46**, 703 (1934)
2.13 R.E.Clay, A.Müller: J. Instr. Elect. Engs. **84**, 261 (1939)
2.14 W.T.Astbury, R.D.Preston: Nature **24**, 460 (1934)
2.15 Z.Nishiyama: J. Japan Met. Soc. **15**, 42 (1940) (in Japanese)
2.16 V.Linnitzki, V.Gorski: Sov. Phys.-Tech. Phys. **3**, 220 (1936)
2.17 R.R.Wilson: Rev. Sci. Instr. **12**, 91 (1941)
2.18 S.Miyake, S.Hoshino: X-sen **8**, 45 (1954) (in Japanese)
2.19 Y.Yoneda, K.Kohra, T.Futagami, M.Koga: Kyushu Univ. Engs. Dept. Rep. **27**, 87 (1954)
2.20 S.Kiyono, M.Kanayama, T.Konno, N.Nagashita: Technol. Rep., Tohoku Univ. **XXVII**, 103 (1936)
2.21 A.Taylor: J. Sci. Instr. **26**, 225 (1949); Rev. Sci. Instr. **27**, 757 (1956)
2.22 D.A.Davies: Rev. Sci. Instr. **30**, 488 (1959)
2.23 P.Gay, P.B.Hirsh, J.S.Thorp, J.N.Keller: Proc. Phys. Soc. (London) B**64**, 374 (1951)
2.24 A.Müller: Proc. Roy. Soc. **117**, 31 (1927)
2.25 A.Müller: Proc. Roy. Soc. **132**, 646 (1931)
2.26 V.E.Cosslett, W.C.Nixon: *X-Ray Microscopy* (Cambridge Univ. Press, London 1960) pp. 220
2.27 W.W.Dolan, W.P.Dyke: Phys. Rev. **95**, 327 (1954)
2.28 A.N.Broers: Appl. Phys. **38**, 1991, 3040 (1967)
2.29 A.Müller: Nature **27**, 128 (1929)
2.30 J.M.W. DuMond, B.B.Watson, B.Hicks: Rev. Sci. Instr. **6**, 183 (1935)
2.31 T. Ishimura, Y.Shiraiwa, M.Sawada: J. Phys. Soc. Japan. **12**, 1064 (1957)
2.32 J.Chikawa: Oyo Butsuri **43**, 230 (1974) (in Japanese)
2.33 W.J.Oosterkamp: Phil. Res. Rep. **3**, 49, 161, 303 (1948)

2.34 T.Sato: Tech. Memo. Oosaka Tech. Univ. **2**, 77 (1956) (in Japanese)
2.35 J.Chikawa, I.Fujimoto: NHK Tech. Monograph **23** (1974)
2.36 M.Green: *X-Ray Optics and X-Ray Microanalysis* (Academic Press, New York, London 1963) pp.185
2.37 A.Guinier: *Théorie et Technique de la Radiocristallographie* (Dunod, Paris 1964)
2.38 A.H.G.Kuntke: Phil. Tech. Rev. **20**, 291 (1958/59)
2.39 Y.Shimura, M.Yoshimatsu, H.Uematsu: 8th International Congress of Crystallography, Collected Abstracts VIII-12 (1969)
2.40 J.R.Pierce: *Theory and Design of Electron Beam*, 2nd ed. (D.Van Nostrand, London 1954)
2.41 D.G.Archard: J. Appl. Phys. **32**, 1505 (1961)
2.42 K.Siegbahn, C.Nordling, G.Johanson, J.Hedman: *ESCA Applied to Free Molecules* (North-Holland, Amsterdam 1969);
      U.Gelius, E.Basilier, S.Svensson, T.Bergmark, K.Siegbahn: *UUIP*-817, Uppsala Univ., Institute of Physics (1973)
2.43 J.Chikawa: Appl. Phys. Letters **13**, 387 (1968)
2.44 S.Kozaki, S.Imai, M.Yoshimatsu: to be submitted to R.S.I.
2.45 R.Shall: Arch. Tech. Messen **74—13**, 117 (1953)
2.46 W.Schaaffs: Z.Angew. Phys. **8**, 299 (1956)
2.47 V.A.Tsukerman, N.I.Zavana, M.A.Manakova: Prib. Tekh. Ekstp. **2**, 434 (1966)
2.48 F.M.Charbonnier: Advan. X-Ray Anal. **15**, 446 (1972)
2.49 J.A.Dantzig, R.E.Green Jr.: Advan. X-Ray Anal. **16** (1973)
2.50 Q.Johnson, R.N.Keeler, J.W.Lyle: Nature **213**, 1114 (1967)
2.51 Q.Johnson, A.C.Mitchell: Phys. Rev. Letters **29**, 1369 (1972)

3. X-Ray Lithography

E. Spiller and R. Feder

With 32 Figures

3.1 Background

X-ray microscopy is a very old technique. The potential of X-rays to produce images of higher resolution than is possible with an optical microscope was immediately recognized after their discovery. However, *Röntgen* realized that the available differences in the refractive index of materials were too small to build an X-ray lens. Credit for the first application of X-rays to microscopy is usually given to *Goby* [3.1]. *Goby* produced a contact shadowgraph of an object with *X*-rays on photographic film and viewed the developed film in an optical microscope. He named his technique X-ray microradiography. The resolutions of the photographic film and the optical system are obvious limitations of *Goby*'s technique. In spite of that, X-ray microscopy has been widely used in biology and medicine, mostly for the differences in penetration and image contrast compared to optical microscopy. Some improvements in the resolution have also been obtained by producing magnified shadowgraphs (projection microradiography [3.2], by using X-ray-sensitive films of higher resolution than photographic film, and by the use of a transmission electron microscope for the magnification [3.3, 4]. There have also been attempts to obtain magnified images directly with X-rays by using grazing incidence mirror optics [3.5]. The resolution of these grazing incidence instruments has not been better than that of an optical microscope due to the aberration of these instruments and to the difficulties in fabricating the desired surfaces. The challenge to build a magnifying X-ray microscope with drastically improved resolution over the optical microscope is still open for future work. Several excellent monograph and review articles on X-ray microscopy are available [3.5–7]. They all contain information which is also useful for X-ray lithography and are highly recommended.

3.2 Description

X-ray lithography is the application of X-ray microscopy to the fabrication of devices like electronic microcircuits. It was established as a viable fabrication technique by *Spears* and *Smith* in 1972 [3.8, 9]. Their first two publications discuss all the essential ingredients required for device fabrication.

1) A mask, consisting of a device pattern made of X-ray absorbing material on a thin membrane of X-ray transmitting material.

2) An X-ray resist of high resolution suitable for the subsequent device fabrication steps as the X-ray sensitive material.

3) An X-ray source of sufficient brightness in the soft X-ray region to expose the resist through the mask.

Figure 3.1 illustrates a simple X-ray lithographic system. X-rays are excited by bombarding a water cooled target material with electrons. In Fig. 3.1 a commercial *e*-beam evaporator[1] with electron energies between 4 and 10 keV is shown as the X-ray source. An absorber pattern on the mask attenuates the incident X-ray intensity; the shadowgraph of the mask pattern modifies the resist film on the device wafer such that in a subsequent development process a replica of the mask pattern in the resist is obtained. Two types of geometrical distortions are illustrated in Fig. 3.1. The first is the penumbral blurring due to the finite size of the X-ray source which limits the resolution to

$$\delta_p = s d_s / D \tag{3.1}$$

where s and D are the mask-wafer and mask source distances and d_s is the diameter of the X-ray source. The second distortion (often called walk-off, run-off or geometrical distortion) is caused by the nonnormal incidence of the X-rays at the edge of the wafer which produces a slightly magnified replica with a displacement δ_m of the pattern on the wafer given by

$$\delta_m = s \tan \Theta \tag{3.2}$$

where Θ is the position-dependent angle of incidence of the X-ray beam on the mask. Both distortions can be controlled by a proper choice of the system geometry.

An electron beam fabrication system [3.10, 11] is used to write the pattern on the mask. It is expected that the subsequent replication of this pattern with X-rays will be much cheaper than the direct electron beam writing of the resist. The main reason is that an electron beam system exposes the pattern point by point and is therefore quite limited in throughput; in addition, electron beam systems are very expensive.

Since the first publication of *Spears* and *Smith*, X-ray lithography has been studied by many workers to explore its capabilities and limits and to develop it into a manufacturing tool. Impressive results have been obtained which in turn could be used to improve the original field of X-ray microscopy.

3.3 Wavelength Selection

Many parameters determine the performance of an X-ray lithographic system and many of these parameters vary with the X-ray wavelength. Examples are the absorption coefficient of the materials used as mask substrate, absorber pattern,

---

[1] Vacuum Generators, Ltd., Model EG-1 or EG-2.

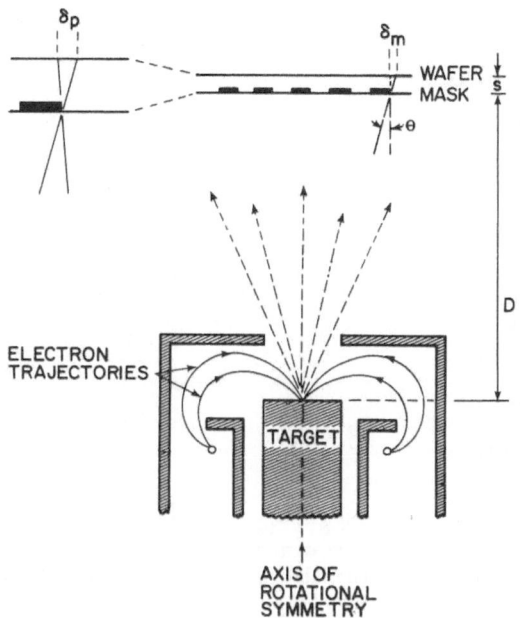

Fig. 3.1. Simple X-ray exposure station. The two geometrical distortions are illustraded: penumbral blurring δ_p due to the source size and the run-off error δ_m due to the wafer size, which results in a slight magnification of the mask

window or resist, the resolution and sensitivity of resists, and the intensity of available sources. There is no optimum wavelength which optimizes a system in every respect; each existing system is a compromise between several contradicting requirements.

Figure 3.2 shows the linear absorption coefficient α of some of the most absorbing and some of the most transparent materials available [3.12–15]. Except for the discontinuities at absorption edges, the absorption increases with increasing wavelengths. The most absorbing materials are only about a factor 50 higher in their absorption constant than the most transparent materials. For comparison a range of about 10^{10} in the values for α is available in the visible spectrum. In the X-ray region, no material is completely transparent and no material is completely opaque. Therefore mask substrates have to be relatively thin to have sufficient transmission and the absorber pattern has to be relatively thick to attenuate X-rays sufficiently. Mask substrates and all windows between source and masks have a higher transmission for shorter wavelength. However, for shorter wavelengths, the absorber material which defines the pattern is also more transparent and the mask has a lower contrast. The selected wavelength is therefore a compromise between the desire to obtain as much transmission and as much contrast as possible. For patterns with dimensions around 1 µm, the wavelength region between 4 and 50 Å represents an acceptable compromise. In this region, light elements have linear absorption coefficients of $\alpha < 1\,\mu m^{-1}$, so that films consisting of light elements with thicknesses of several µm can be used as a mask substrate and give acceptable transmission values. The heavy elements have absorption coefficients of $\alpha > 1\,\mu m^{-1}$, such that absorber patterns with

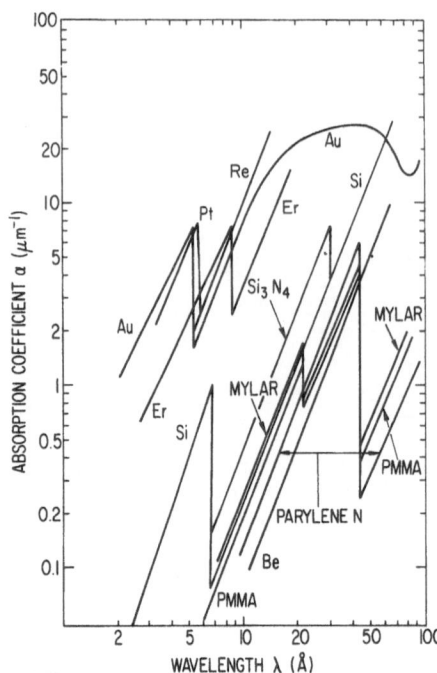

Fig. 3.2. Absorption coefficients for some of the strongest and some of the weakest absorber materials in the soft X-ray region. PMMA is a widely used X-ray resist (see Sec. 3.6)

thickness below one micrometer give acceptable contrast. It is clear from the absorption curves in Fig. 3.2 that the shorter wavelengths (around $\lambda = 5$ Å) are most useful for relatively coarse structures which allow relatively thick absorber patterns, while for the replication of high resolution patterns (linewidth < 0.1 μm) long wavelengths (around $\lambda = 50$ Å) are to be preferred.

We will show in Section 3.6 that, in addition, resist materials have a better resolution at longer wavelength as long as diffraction effects can be neglected. Diffraction effects can become important for wavelengths above $\lambda = 50$ Å, where resolution is limited by diffraction and not by properties of the resist material. The wavelength region around $\lambda = 50$ Å therefore gives the highest potential resolution and simultaneously the best contrast.

Figure 3.2 also gives the absorption curve for one widely used resist material: polymethylmethacrylate (PMMA). The higher absorption coefficient of the resist at longer wavelength is equivalent to a higher sensitivity of the resist at longer wavelength because a larger fraction of the incident radiation is absorbed in a thin resist layer. This higher sensitivity can partly compensate for the higher absorption losses in the substrate such that exposure times are not necessarily longer for longer wavelengths (provided that $\alpha d < 1$ for both the resist film and the substrate, $d =$ thickness).

However, in general, thinner mask substrates are required for longer wavelengths, and the use of vacuum windows to separate the X-ray source and the exposure station for exposure at atmospheric pressure is very limited.

Exposure in an atmosphere of He is no problem for systems operating around wavelengths of 5 Å; resist sensitivity can be increased in this region by doping the resist with a heavy material to increase its absorption (see Sect. 3.6).

The convenience associated with thicker substrates and the use of a vacuum window with a possible exposure in a nonvacuum environment make shorter wavelengths ($\lambda \approx 5$ Å) preferable for relatively coarse structures with linewidths larger than 1 µm. For the finest structures, the need for high resolution and sufficient contrast requires the use of longer wavelengths ($\lambda = 50$ Å); at these longer wavelengths the mask substrate has to be very much thinner for sufficient transmission and the exposure has to be made in a vacuum chamber.

3.4 Masks

Any material transparent to X-rays can be used as a substrate for an X-ray mask. In addition to providing high transmission, an X-ray mask substrate should be flat, smooth, self-supporting, dimensionally stable, preferably optically transparent, and sufficiently rugged to withstand the processing steps required for the mask fabrication and the handling during its use. Table 3.1 gives a listing of some light materials which can be obtained in thin film form and can be used as a mask substrate. One can determine the wavelength range over which each substrate in Table 3.1 is usable from the given thickness ranges and the absorption data in Fig. 3.2.

Any heavy material with a sufficiently high absorption can be used to define the opaque areas of the mask. Gold has been reported as the preferred absorber material by practically everybody, mainly because its properties in all subsequent processing steps are well known, and much higher absorption values cannot be expected with other materials.

Despite the fact that an X-ray mask is a very simple structure (an X-ray absorbing pattern defined on an X-ray transparent thin substrate), there are numerous combinations of methods to fabricate such a mask depending on the process steps chosen to define the pattern and to produce the thin film substrate. Every processing step listed in Section 3.8 has probably been used to define the absorber pattern in an X-ray mask. Another variation is the order in which the mask substrate and the absorber pattern are fabricated. The thin substrate can be fabricated first and then the absorber pattern can be defined on top of this substrate film. Alternatively, the absorber pattern can be fabricated first on a thick support wafer which is thinned out later, or individual processing steps for the pattern fabrication and for the substrate fabrication can follow each other in any sequence which might appear advantageous for the selected processes and materials.

In the following we shall discuss some of the mask fabrication steps which have been used for different mask substrates.

Table 3.1 Substrate materials for X-ray masks or windows

Substrate	Thickness [μm]	Wavelength region [Å]	References
Silicon ($p+$)	1–5	7–15	[3.8, 9, 16–20]
Beryllium	12		[3.8, 9, 16–20]
Mylar[a] (Hostaphan[b])	2–25	<20 44–60	[3.18, 22, 23]
Polyimide (Kapton[a, c])	2–25	<20 44–60	[3.22, 24, 25, 28]
Si_3N_4	0.1–0.2	<40	[3.26, 27]
$Si_3N_4 + SiO_2$	0.3–0.4	<20	[3.27]
Al_2O_3	3–5	8–15	[3.29, 36]
Polypropylene	1–25	<25 44–70	[3.30, 35]
Polycarbonate[d] (Kimfol[e])	2	<20 44–60	[3.30]
Polyvinyl formal[d] (Formvar)	0.4	<30 44–75	[3.30]
Aluminium[d]	6.4	8–14	[3.18, 30]
Nitrocellulose[f]		<20 44–60	[3.31, 32]
Parylene[d]	0.2–3	<70	[3.33]

[a] Trademark of Dupont, Wilmington, Delaware.
[b] Trademark of Kalle, Wiesbaden, Germany.
[c] Also Monsanto Skybond 704.
[d] Used as soft X-ray window only.
[e] Trademark of Kimberly-Clark, Lee, Mass.
[f] Available from National Photocolor Corp., Westport, CT.

3.4.1 Silicon Membranes

Silicon membranes have been used as X-ray mask by *Spears* and *Smith* in their first publication [3.8]. Figure 3.3 shows one possible sequence of fabrication steps to obtain such a mask. The key in the fabrication of a silicon membrane is the fact that a solution of ethylene diamine, pyrocatechol, and water at 115 °C and also diluted HF solutions etch all types of silicon except silicon that is very heavily doped with boron. In the sequence of Fig. 3.3 the absorber pattern is first fabricated on a silicon wafer which contains a heavy boron doping at the top surface (about 5 μm thick). Afterwards the silicon is etched from the back side through windows in a protective SiO_2 layer leaving the thin boron doped membrane stretched over the frame of silicon which was protected during the etching. The finished membranes are under tension and are very flat if the support framework is made thick enough to give sufficient rigidity. The membranes are transparent for red light. Si membranes have been produced over areas up to 4 cm diameter and with thicknesses between 1 and 5 μm [3.16–21]. They are useful for the wavelength region $\lambda = 7$–15 Å.

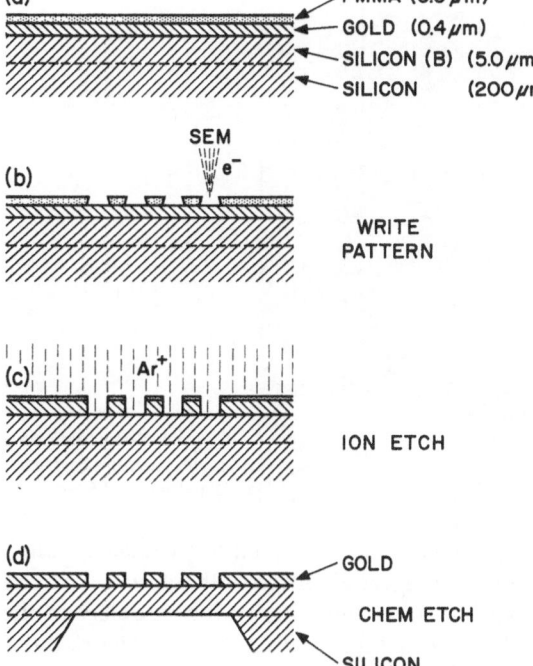

Fig. 3.3. Fabrication steps for an X-ray mask on a silicon membrane with gold absorber pattern (courtesy of *Smith* [3.16])

3.4.2 Si_3N_4 and $Si_3N_4 + SiO_2$ Membranes

Si_3N_4 and $Si_3N_4 + SiO_2$ membranes are fabricated on silicon wafers in the same way as Si-membranes. The doping of the top of the Si wafer is replaced by a chemical vapor deposition of Si_3N_4 or $Si_3N_4 + SiO_2$; afterwards the same steps as for Si membranes can be used [3.8, 26, 27]. The main advantages of the $Si_3N_4 + SiO_2$ membranes over silicon membranes are: The membranes can be made very much thinner ($d = 0.1$–0.4 μm) and they are completely transparent to visible light. The membranes have been produced in windows of 2 mm × 2 mm size. They are useful for soft X-rays up to $\lambda = 40$ Å and have been used for high resolution lithography with Cu L radiation with $\lambda = 13$ Å [3.26].

3.4.3 Organic Films

Many organic films can be prepared and are commercially available in thin film form. The mask fabrication steps are drastically simplified with commercially available films which have only to be mounted on a frame and are ready for the processing to define the absorber pattern (Fig. 3.4). Because of this convenience, Mylar and Hostaphan have been widely used as substrates for X-ray masks. They have, however, disadvantages for high quality device fabrication where a defect-free lithography is required. During the fabrication of thin mylar films,

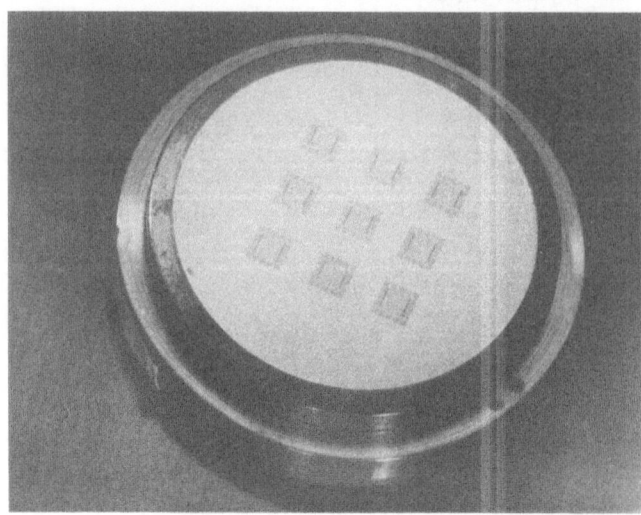

Fig. 3.4. X-ray mask on a mylar membrane, with a diameter of 2.5 cm

small dust particles are deliberately added to the material to produce a rough surface which allows easier rolling of the material. Some of these dust particles contain heavy elements which absorb X-rays and can produce a defect in the replica. Therefore, Mylar is an ideal substrate for initial experimental work but is usually not acceptable for large area device fabrication.

Polyimide is available commercially in sheet form (Kapton); it is also available in solution, and one can fabricate and polymerize the material as a thin film to one's own standards [3.28]. Polyimide can tolerate the highest temperatures of all polymers and allows, therefore, the widest freedom in the selection of the subsequent processing steps.

Nitrocellulose [3.32] and Parylene [3.33] are commercially available already mounted on a frame. However, these materials do not allow further processing at temperatures above 100 °C and have not been used extensively for this reason.

All organic films have a transmission window for X-ray wavelengths above the carbon K_α absorption edge ($\lambda = 44.8$ Å) and are therefore very well suited for extremely high resolution work at such long wavelengths (see Sect. 3.6). In addition, they can be used at shorter wavelengths ($\lambda < 20$ Å).

3.5 X-Ray Sources

3.5.1 X-Ray Generation by Electron Bombardment

Most commercial X-ray tubes are not suitable for X-ray lithography because the emitted spectrum contains radiation which is too hard for sufficient contrast of the mask and for sufficient absorption and sensitivity of the resist. Some early work in X-ray lithography [3.34] was severely limited by the use of the then-available commercial X-ray sources with wavelengths of $\lambda = 2.3$ and $\lambda = 1.54$ Å.

Commercial X-ray systems recently designed for X-ray lithography are discussed by *Yoshimatsu* and *Kozaki* in Chapter 2.

After *Spears* and *Smith* [3.9, 16] showed that a commercial electron beam evaporation source (such as Vacuum Generators, Ltd., Model EG-1 or EG-2) operated at reduced power to avoid evaporation, makes an excellent X-ray source for X-ray lithography, this unit became very popular. A diagram of this source is given in Fig. 3.1. The electron gun has rotational symmetry and the electrons emitted from a circular filament are electrostatically focused on the water-cooled target on the axis of symmetry. The electrons strike the target with a large aperture angle thus minimizing space charge effects allowing for a small spot size (about 1 mm) even at very low accelerating voltages. The position of the best focus is independent of voltage. The possible range of voltages (3–10 kV) is suited for the excitation of soft X-rays in the 5–50 Å wavelength region. Target materials can easily be changed to produce characteristic lines of different elements. Reference [3.30] gives a listing of soft emission lines. In X-ray lithography, Al K_α radiation ($\lambda = 8.3$ Å) has been most often used; other target materials include Rh ($\lambda = 4.6$ Å [3.23]), Pd ($\lambda = 4.37$ Å [3.23, 37, 38]), Cu ($\lambda = 13.3$ Å [3.26]), and C ($\lambda = 44.8$ Å [3.39, 40]). For the generation of very soft X-rays, contamination and roughening of the target is often a serious problem because it can drastically decrease the intensity of the wanted characteristic line. A short surge in the current of the electron gun which melts the target material can be used to provide a fresh surface with the full X-ray intensity. This method works very well for the widely used aluminum targets.

The X-ray source of Fig. 3.1 can be operated up to power levels around 500 W for a 1 mm spot size. For still higher power levels, a rotating water cooled anode is required. Figure 3.5 shows a source which can be considered as a scaled-up version of the evaporator gun with similar properties but with a power capability of about 10 kW [3.35]. Separate electrodes for Wehnelt, acceleration, and deflection allow the focus to be adjusted electrically from the outside. In addition, voltage, emission current, and filament current can be adjusted independently while in the original evaporation gun the current is space charge limited. Commercial X-ray generators use straight optical axis Pierce-type guns (see Chap. 2). These guns are easier to fabricate than ring cathode systems. For operation at low accelerating voltages ($V < 10$ kV), Pierce-type guns give larger spot sizes and less brightness than the ring cathode guns due to space charge effects close to the beam focus. At higher voltages, space charge effects become irrelevant and Pierce-type guns are preferred to ring cathode guns. The energy efficiency for the generation of characteristic X-rays increases initially with electron energy above the threshold energy. For Al K_α radiation the highest efficiency is reached for an acceleration voltage of ~ 15 kV, for carbon radiation 5 kV is an optimum voltage [3.36, 41]. Continuous Bremsstrahlung is also generated more efficiently and extends to higher energies with increasing voltages [3.16, 24, 41]. The harder continuous radiation is usually less absorbed by the absorber pattern resulting in a lower effective contrast of the mask. *Because most resists also have a lower sensitivity for harder radiation, the effect*

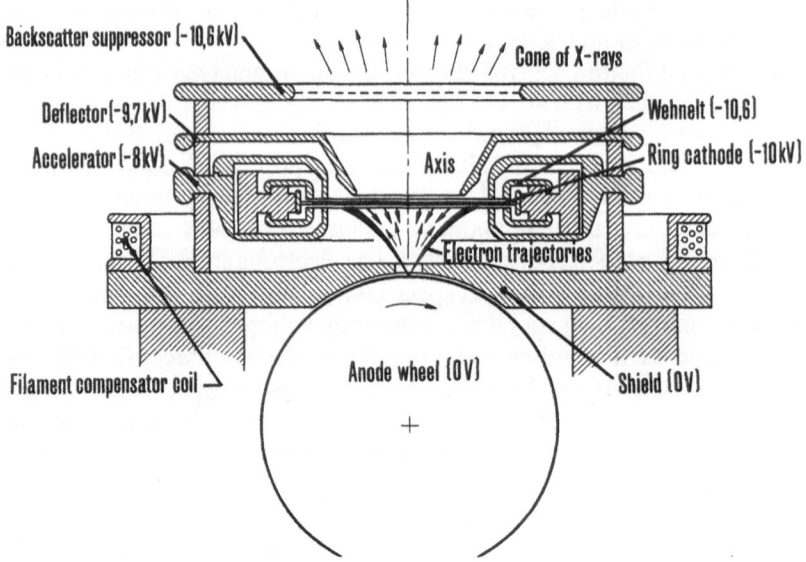

Fig. 3.5. Rotating anode high power X-ray source for soft X-rays (courtesy of *Wardly* [3.35])

of the Bremsstrahlung on the contrast is not as strong as one might expect from the absorption curves alone. *Sullivan* and *McCoy* [3.41] have given the effective contrast of gold absorber patterns as a function of the electron beam energy for an aluminum target and several mask-substrate and window combinations, while *Maldonado* et al. [3.42] have published similar data for a Pd source with 20 kV acceleration voltage. The contrast deterioration due to insufficient absorption of the mask absorber pattern, however, is not the only problem caused by the harder continuous radiation. In many cases, the secondary electrons generated by absorption of the incident radiation in the Au layer of the mask or in the substrate have more serious effects. These secondary electrons will expose the surrounding resist and in some cases give a heavier exposure for areas under the absorber pattern than for clear areas [3.43]. The secondary electrons emitted from the substrate back into the resist make the development process very critical; poor adhesion of the resist due to the additional exposure can generate problems for the subsequent processing steps [3.42–44].

For these reasons it is often advantageous to operate the X-ray source at a lower voltage than is optimum for the most efficient energy conversion.

The common disadvantage of all X-ray sources which excite X-rays by bombarding a target with electrons is the low conversion efficiency from electron beam power to X-ray power ($\eta < 10^{-4}$). A high power 10 kW rotating anode gun delivers at most an X-ray output of only 1 W into a large solid angle and therefore the efficient use of this X-ray flux is an important consideration in the design of every lithographic system.

3.5.2 Synchrotron Radiation Sources

Synchrotron radiation sources are by far the brightest sources of soft X-rays [3.45–49]. Synchrotron radiation is emitted by high energy relativistic electrons in a synchrotron or storage ring, which are accelerated normal to the direction of their motion by a magnetic field (Fig. 3.6). The spectrum of the emitted radiation extends from the microwave region throughout the infrared, visible, ultraviolet and into the X-ray region, decreasing in intensity below a critical X-ray wavelength

$$\lambda_c[\text{Å}] = 187/B[\text{kOe}]E^2[\text{GeV}], \tag{3.3}$$

where E is the energy of the electrons in GeV and B is the magnetic flux in kOe. The radiation is highly collimated in the direction of the instantaneous electron motion. Thus, as the electrons traverse the magnetic field, the emitted radiation sweeps in an arc in the plane of the electron orbit. In the vertical direction the beam has an angular spread which can be approximated by

$$\psi = \frac{mc^2}{E} \cdot \left(\frac{\lambda}{\lambda_c}\right)^{1/3}, \tag{3.4}$$

where E is the electron energy and $mc^2 = 0.511$ MeV is the electron rest mass. For a 1 GeV electron we obtain an angular spread of about 0.5 mrad at λ_c.

One can understand the emitted radiation as dipole radiation generated by the acceleration of the electron perpendicular to its motion, which has been Doppler shifted towards very high frequencies in the forward direction and to a frequency close to zero in the backward direction due to the relativistic speed of the electrons.

The small angular spread of the beam in the vertical direction does not allow exposures of large areas simultaneously; in addition, intensity and exposure spectrum depend on heights. For uniform exposure over large areas, one would move the mask wafer combination with constant speed through the fan of synchrotron radiation (Fig. 3.6).

Synchrotron radiation is presently available in several laboratories, where synchrotron or storage rings have been built primarily for experiments in high energy physics and where the emitted synchrotron radiation is used in a parasitic mode [3.45–48]. The idea to build a storage ring solely as a radiation source has gained more and more support in recent years and several of these projects are now planned or are under construction. The few storage rings already in operation as dedicated sources [3.47] have electron energies which are too low, so that the emitted spectrum is not usable for X-ray lithography.

Electron storage rings designed as a light source for X-ray lithography would have an electron energy around 1 GeV with a wavelength spectrum extending below $\lambda = 10$ Å for magnetic fields around 20 kOe. Some superconducting bending magnets with higher magnetic fields could be provided for the experiments which require shorter wavelengths. For mask replication with

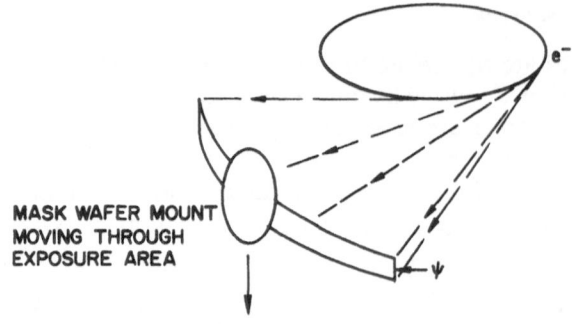

Fig. 3.6. Schematic of an X-ray exposure station with a synchrotron radiation source

MASK WAFER MOUNT MOVING THROUGH EXPOSURE AREA

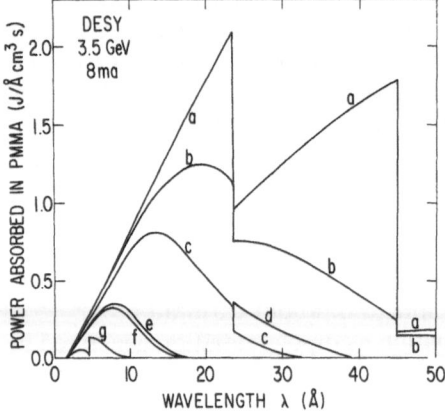

Fig. 3.7. Power absorbed in resist (PMMA) as a function of wavelength for synchrotron radiation from DESY (3.5 GeV, 8 mA) 40 m from electron orbit for various filters in the beam. a no filter, b 0.3 μm of PMMA, c 1.5 μm of Parylene N, d 1 μm of PMMA, e 12 μm of beryllium, f 6 μm of mylar, g 6 μm of mylar $+0.8$ μm of gold

extreme resolution requirements and for high resolution X-ray microscopy, beams with only very soft X-rays could be made available either at sections with very low magnetic fields or by lowering the energy of the electrons. A machine with much higher energy than 1 GeV has practically no advantages for lithographic applications. Radiation safety rules will make the use of such a machine much less convenient (remotely controlled experiments) and only add drastically to the costs. The main disadvantage of synchrotron radiation sources is their cost. Although dedicated sources can be considerably smaller and cheaper than the machines built for high energy physics, the cost is still several million dollars [3.49]. A dedicated storage ring used for X-ray lithography would have a throughput capability which is several orders of magnitude higher than all projected production requirements. Because the cost of the machine is mainly determined by λ_c, which determines the size of the machine and of the magnets, not much can be saved by reducing the maximum electron current for a lower output intensity.

The DESY electron sychrotron in Hamburg and the ACO storage ring have been used for X-ray lithography [3.50, 51, 53, 54] and X-ray microscopy [3.40, 52]. Figure 3.7 shows the exposure spectrum of an X-ray resist for the DESY synchrotron operating at an electron energy of 3.5 GeV. The different curves are

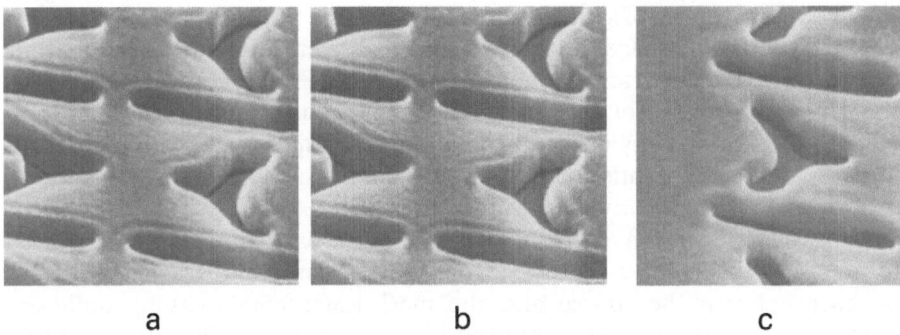

a   b   c

Fig. 3.8a–c. Mask replication (1 μm linewidth) with various distances between mask and wafer: Filter: 12 μm Be + 6 μm mylar; exposure time: 3 min; effective exposure 250 J cm^{-3}; resist: PMMA 2041; mask-wafer spacing (a) 0.14 mm, (b) 0.54 mm, (c) 1.04 mm, synchrotron radiation from DESY at 3.5 GeV electron energy

for different filter materials, which can represent the substrate, the absorber, an extra filter, or the top layer of resist which acts as a spectral filter for the deeper resist layers. Curve f shows the exposure spectrum of a 6 μm thick layer of a polymer film (mylar), which was used as a mask substrate; curve g gives the exposure spectrum for a 0.8 μm thick gold film on that substrate; the ratio of the areas under the curves f and g determines the effective contrast of the mask. The curves show that short wavelengths in the DESY beam reduce the contrast of the mask considerably and that a lower energy of the electrons in the synchrotron with a wavelength cutoff around 10 Å would improve the contrast. Hard radiation cannot be easily eliminated with a filter (see Fig. 3.2); reflection from a mirror with a small glancing angle of 4° was used to produce a spectrum in the 30–50 Å wavelength range suitable for the highest resolution [3.40, 50]. The ACO storage ring in Orsay has a maximum operating energy of 540 MeV; the emitted spectrum which is peaked at $\lambda = 20$ Å is very suitable for high resolution work.

Figure 3.8 demonstrates the excellent collimation of the DESY synchrotron. Penumbral blurring is completely eliminated and one can tolerate large distances between mask and wafer without loss in resolution. However, at very large distances, diffraction effects determine the resolution limit; these diffraction effects are visible in Fig. 3.8b and c for mask wafer distances of 0.54 and 1.04 mm.

3.5.3 Laser Generated Plasmas as X-Ray Sources

The hot plasma which is generated in the focus of a high power laser represents a bright source of X-rays. A neodymium laser with a pulse energy of 100 J, developed at Battelle's Columbus Laboratories [3.153, 154] produces about 30 J of soft X-rays generated in a plasma with dimensions between 100 and 200 μm within a pulselength of 5 n. This X-ray flux is more than sufficient to expose even the least sensitive resist materials.

The main problem of a lithographic system using the laser generated plasma as X-ray source is damage to the mask and the wafer. The short exposure time does not allow dissipation of the energy absorbed in mask and wafer during the exposure. A resist of very high resolution requires an exposure close to 1 J cm^{-2} and this energy is absorbed within a thickness of about 10 μm resulting in an energy absorption close to 1000 J cm^{-3}. Such an energy density will destroy any resist and any mask substrate.

First experiments have shown [3.155] that the exposure has to be reduced to about 1 mJ cm^{-2} per pulse to avoid mask destruction (by increasing the distance between the source and the mask-wafer combination). Sufficient exposure of the resist is then obtained by using several pulses (pulse spacing 5 min). For a high sensitivity resist like DClPA (see Table 3.2) a usable exposure has been obtained with four pulses. One can hope that more sensitive resists which allow exposure in one pulse will be available in the future. With such high-sensitivity resists laser generated plasmas might become very useful sources for X-ray lithographic systems; the required resist sensitivity will, however, limit the resolution for single pulse exposure.

3.6 X-Ray Resists

All resist materials in practical use are films of organic polymers. In a first step the device wafer is coated with the resist film (in most cases by spin coating, sometimes by dipping the wafer into a solution of resist) and then dried (baked). The second step is the exposure of the resist with the wanted pattern; in this exposure the resist is modified by the radiation in such a way that a subsequent development process can distinguish between unexposed and exposed regions, removing one and leaving the other intact. The remaining resist then serves as a protective coating for the subsequent processing. There have been efforts to simplify the procedure by modifying the device wafer directly by the radiation (e.g., change the etch rate of SiO_2 by electron bombardment) and selecting subsequent processing steps which take advantage of these modifications [3.55–58]; however, these efforts have not led to any process which is used in practice and therefore will not be described here.

All processes in practical use today use wet development: the wafer with the exposed resist is immersed in or sprayed with a weak solvent which removes exposed and unexposed resist regions at a different rate. Some cases have been reported in which the resist film was removed without any wet development either directly during exposure or by some subsequent dry treatment like heating [3.59–63]. These processes are also not used in practice today and will not be described here.

We will describe resist materials from a user's point of view. Parameters like sensitivity, contrast, and resolution which are important for the selection of a resist for a specific application will be discussed. The details of the chemistry of the resist will be omitted, i.e., which chemical structures produce high sensitivity, high contrast or high resolution, which structures give good resistance for subsequent processing steps, produce good uniform films, etc. This knowledge is important for the chemist who wants to design a new resist but is of only secondary importance for the user of a given resist. The reader interested in the chemistry is referred to the literature [3.64–68]. Designing a new resist is still more of an art than a science. Once a new resist has been made, explanations for its behavior can usually be derived from its structure; very rarely does a new resist behave exactly the way it was predicted before the material was synthesized.

The exposure of polymer films in electron beam and in X-ray lithography is a very similar process. The energy of the incident particles is in both cases much larger than the energy required to form or to break a chemical bond. X-rays have a much weaker interaction with the resist than do electrons. Once an X-ray photon is absorbed it produces a shower of secondary electrons which will carry most of the energy of the incident photon and then have a much stronger interaction with the resist material. Therefore, these secondary electrons are practically responsible for all the chemical changes in the resist [3.8]. X-ray exposure of a resist is in essence an exposure by electrons; the only difference between an X-ray and an electron beam exposure system is that in an electron beam system the incident electrons have an energy in the range between 10 and 50 keV, while the secondary electrons produced by soft X-ray exposure have much smaller energies.

Therefore, all electron beam resists are also usable as X-ray resists. The same energy deposition in a volume element of resist will, in a first approximation, produce the same changes for an electron or an X-ray exposure. Nearly all work on electron beam resists is also useful for X-ray resists and vice versa.

3.6.1 Performance of an Ideal Resist

The shot noise in the absorption of photons defines a limitation for the sensitivity of any detector. An ideal resist is a resist film which has no other limitations in its performance. It is worthwhile to discuss the performance of such an ideal resist. Goals for research work on resists can be better defined by comparing the performance of existing resists to the ultimate performance of the ideal resist.

The average number of X-ray photons absorbed in a small volume of resist (Fig. 3.9) is given by

$$\bar{n} = \frac{E_{inc}}{hv} \cdot A \cdot \delta^2 , \qquad (3.5)$$

where E_{inc} is the incident exposure density (e.g., in J cm^{-2}), hv is the energy of each X-ray photon, and A is the fraction of the incident energy which is absorbed in the resolution element of resist with lateral dimension δ and a thickness d. The absorption is given by

$$A = 1 - e^{-\alpha d} , \qquad (3.6a)$$

$$A \approx \alpha d \quad \text{for} \quad \alpha d \ll 1 \qquad (3.6b)$$

with the linear absorption coefficient α.

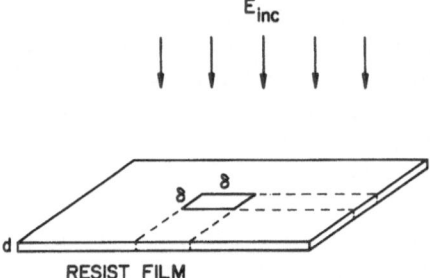

E_{inc}

RESIST FILM

Fig. 3.9. Resolution element of a resist film

Because photons arrive at random and independently from each other, the actual number of photons absorbed in different resolution elements will deviate from the average value. These deviations from the average value, are responsible for the performance limitations of an ideal resist. The probability of absorbing photons in one resolution element is determined by the Poisson distribution [3.69]

$$p(n, \bar{n}) = \frac{\bar{n}^n}{n!} e^{-\bar{n}}, \qquad (3.7a)$$

which for large $\bar{n}$ can be approximated by

$$p(n, \bar{n}) \approx \frac{1}{\sqrt{2\pi\bar{n}}} \exp\left[\frac{-(n-\bar{n})^2}{2\bar{n}}\right]. \qquad (3.7b)$$

The Poisson distribution has a width defined by its standard deviation

$$\sigma = \sqrt{\bar{n}}. \qquad (3.8)$$

An X-ray mask consists of areas of different transmission values. Let us assume that the transmission is 1 in the "clear" part and has the value T_{abs} in the "opaque" parts which contain the absorber material. The average exposure in the opaque part is then reduced to $\bar{n}_{\text{abs}} = T_{\text{abs}}\,\bar{n}$ as compared to the value $\bar{n}$ in the clear parts of the mask.

The probability $p(n)$ to observe n photons in a resolution element is plotted in Fig. 3.10 for areas behind the clear and the opaque parts of the masks under the assumption that the average $\bar{n} = 15$ in the clear part and $\bar{n} = 7.5$ in the opaque part ($T_{\text{abs}} = 0.5$). For the example of Fig. 3.10 the two Poisson distributions overlap. We have no way of knowing from the exposed resist copy whether the mask was transparent or opaque in a special area for observed photon absorptions which are in the overlap region (around $n = 11$ in Fig. 3.10). For an ideal resist we might define an optimum threshold value n_{opt} such that all elements which received $n \leqq n_{\text{opt}}$ photons are developing as if not exposed and all resist areas with $n > n_{\text{opt}}$ are recognized as exposed areas and choose n_{opt} in such a way as to minimize the probability for an error. Whenever we observe $n \leqq n_{\text{opt}}$ photons in the clear sections in the mask or $n > n_{\text{opt}}$ photons in the opaque sections, we have made an error and produced a defect. The probability for such a defect is

$$P_{\text{def}} = \sum_{n=0}^{n_{\text{opt}}} p(n, \bar{n}) + \sum_{n=n_{\text{opt}}+1}^{\infty} p(n, T_{\text{abs}} \cdot \bar{n}). \qquad (3.9)$$

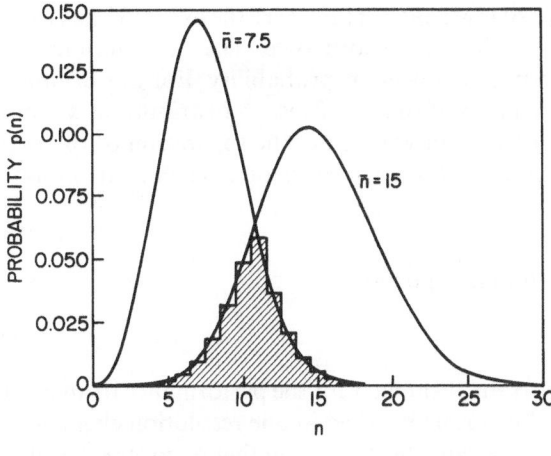

Fig. 3.10. Poisson distributions for an exposure $\bar{n}=15$ in the clear and $\bar{n}=7.5$ in the opaque ($T_{abs}=0.5$) areas of the mask. The shaded area represents the probability for a defect, which occurs when more than $n_{opt}=11$ photons are observed for the opaque areas or $n_{opt}=11$ or less photons for the clear areas of the mask

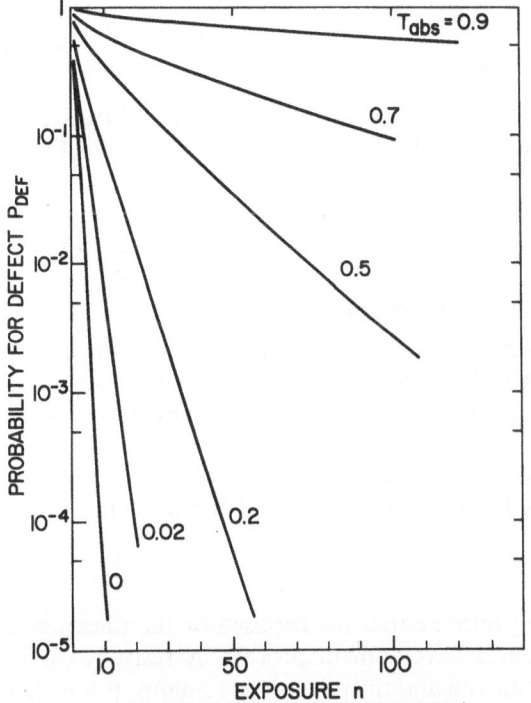

Fig. 3.11. Probability for a defect P_{def} as a function of exposure in the clear areas of a mask for an ideal resist for different transmission values in the opaque areas of a mask

In Fig. 3.10, P_{def} is represented by the shaded area under the two Poisson distributions. The defect probability becomes smaller for a larger separation of the two Poisson distributions which can be achieved by increasing the exposure and/or by increasing the contrast of the mask ($C_{mask}=1/T_{abs}$).

Figure 3.11 is a plot of P_{def} as a function of the exposure for various transmission values of the absorber pattern. For high contrast masks ($T_{abs}<0.02$), practically defect-free resist copies ($P_{def}<10^{-6}$) can be obtained with exposures $\bar{n} > 30$. For masks of low contrast ($T_{abs} \approx 1$), the exposure has to be increased drastically to obtain a similar performance. The 3σ-criterion is a widely used measure for an acceptable low error probability. For our example this criterion requires that the overlap of the two Poisson distributions should only occur outside the 3σ limit of each distribution. The separation of the two distributions should be three times the standard deviation of each distribution:

$$\bar{n} - T_{abs} \cdot \bar{n} = 3\sigma_{clear} + 3\sigma_{opaque} . \tag{3.10}$$

With (3.8) we obtain for the required exposure

$$\bar{n} = 9/(1 - \sqrt{T_{abs}})^2 . \tag{3.11}$$

We have shown that the important parameter for the performance of an ideal resist is the average number of photons $\bar{n}$ absorbed in one resolution element of the resist. We can rewrite (3.5) to obtain the corresponding exposure density

$$E_{inc} = \frac{\bar{n} \cdot h\nu}{A \cdot \delta^2} . \tag{3.12a}$$

For extremely high resolution ($\delta \to \lambda$) the resist has to be very thin ($d \approx \delta$) and therefore absorbs only a small fraction of the incident radiation. Inserting (3.6b) in (3.12) with $d = \delta$, we obtain

$$E_{inc} = \frac{\bar{n} \cdot h\nu}{\alpha \cdot \delta^3} . \tag{3.12b}$$

The ideal resist is characterized by the following properties:

1) The required exposure increases at least quadratically with the linear resolution (3.12). High resolution resists must have low sensitivity.

2) For photons of higher energy (shorter wavelength) higher exposures are required (3.12).

3) Low contrast masks require higher exposures (3.11). For high sensitivity the resist has a lower contrast and requires masks of higher contrast.

3.6.2 Real Resists

All resists belong to one of two groups: either the exposed or the unexposed areas of the resist are removed in a development process. A resist is called "positive" if the exposed area is removed and the unexposed remains. It is called "negative" if the unexposed area is removed and the exposed area remains. Traditionally, resists have been used with wet acid etching as the subsequent processing step, where the remaining resist (resist = resistant to acid) protected an underlying surface from the etching of the acid. With this processing step a positive resist produced a positive copy of a metal mask while a negative resist reversed the polarity and produced a negative. Today wet etching is more and more replaced by other processes, which may be subtractive like acid etching or additive (material is added where the resist has been removed) like electroplating. Positive and negative copies of a mask can therefore be obtained with the same resist, depending on the choice of the subsequent processing steps. The traditional classification of resists has lost its original meaning but is still used.

Once the compatibility of a resist with all processing steps has been established, it has to be characterized by its lithographic radiation sensitivity, its contrast, and its resolution. From the behavior of the ideal resist we might expect that these parameters are connected. In addition, the requirements on resist performance depend strongly on the processing steps used, and different definitions of sensitivity and contrast are preferred for different processes. For these reasons there exists no generally accepted rule how to characterize a resist. Values for sensitivity and contrast given in the literature for different resists can only be compared with caution, because different characterization methods may have been used.

Resist Characterization (Example PMMA)

We shall use polymethylmethacrylate resist (PMMA) as an example to discuss the various methods of characterizing a resist. PMMA as a resist was discovered by *Haller* et al. [3.70] and has been used extensively as an *e*-beam resist for the past 10 years. X-ray lithography became a viable technique when *Spears* and *Smith* [3.8, 9] demonstrated its capabilities with PMMA as the resist. PMMA has the highest resolution of all existing resists and has a very high contrast. Its sensitivity, however, is lower than that of any other X-ray resist which is in practical use.

The dissolution rate S of PMMA in a solvent increases if the resist is exposed to X-rays or electrons; it is, therefore a positive resist. The chemical changes induced by the radiation have been discussed by *Hiraoka* [3.64]. A reduction of the molecular weight of the polymer by chain scission is the most important modification of the resist. The dissolution rate S in a solvent changes with molecular weight $M = N M_m$ according to

$$S \propto M^{-a} \tag{3.13}$$

with a value $a \approx 1.5$ [3.71, 72] for PMMA (Dupont 2041) developed in concentrated methyl isobutyl ketone (MIBK) at 21 °C. M_m is the molecular weight of a mononer unit and N the average number of mononer units per chain. It follows from (3.13) that we can expect larger changes in the dissolution rate after exposure for resists of higher molecular weight. The breaking of one bond out of the N bonds in a polymer chain can reduce the molecular weight to one half of its initial value with the corresponding increase in the dissolution rate given by (3.13). For larger N, this one broken bond represents a smaller fraction of the total number of bonds per volume and less exposure is required. The use of higher molecular weight material therefore seems an obvious way to increase the sensitivity of a polymer resist. Practical reasons limit this approach. It is difficult to prepare extremely high molecular weight materials with a narrow distribution of the molecular weights; they cannot be easily dissolved, even in strong solvents, and are therefore inconvenient to use. For these reasons the highest molecular weights used are around 10^6.

Dissolution Rate. Figure 3.12 gives the dissolution rate S of PMMA as a function of exposure for different developer solutions. At very low exposures the resist dissolves very slowly and we use a strong developer to develop the resist in a reasonably short time (methyl isobutyl ketone + 1 percent water). Unexposed resist is removed in this developer at a rate $S_0 \approx (70 \pm 25)$ Å min⁻¹. Exposure increases the dissolution rate, first slowly and then faster and faster. For an exposure of 10^3 J cm⁻³ the dissolution rate has increased 10^3 times over the initial dissolution rate; a 1 μm thick resist film is removed within a few seconds. At still higher exposures the required development time to remove the resist completely becomes too short to be controllable. Developers diluted with isopropyl-alcohol give usable development times (in the order of minutes) for these high exposures. For exposures above 10^4 J cm⁻³ the resist is dissolved very fast even in isopropyl-alcohol. However, a reversion of the trend occurs for exposures above 5×10^4 J cm⁻³. Some insoluble residue is first observed and at 10^5 J cm⁻³ the exposed resist is completely insoluble even in concentrated MIBK. The resist behaves as a negative resist, the unexposed resist is now dissolved away. (The dissolution rate of the "unexposed" resist at 10^5 J cm⁻³ exposure is higher than S_0 because the mask used for this exposure was not completely opaque.) It appears that the crosslinking of the resist has produced a three-dimensional insoluble network at these high exposures. In most polymers, chain scission and crosslinking can be induced simultaneously by ionizing radiation [3.67, 70]. For PMMA, chain scission dominates for low exposures; at very high exposures crosslinking prevails [3.73].

The curves in Fig. 3.12 were plotted from data obtained from different X-ray sources. Values for low exposures have been obtained with Al K_α radiation ($\lambda = 8.3$ Å) [3.9, 22]. For the very high exposures, synchrotron radiation with a wide wavelength spectrum was used [3.50]. The two exposure scales in Fig. 3.12 were obtained with the assumption that all X-ray wavelengths have the same effect if the same amount of energy is deposited per volume element of resist. No indication that this assumption might be wrong has been observed up to now; however, no efforts have been made to find results which disagree with this assumption. The exposure scale from [3.22] is shifted by around a factor 2 in Fig. 3.12 to be compatible with the results from synchrotron radiation experiments. It appears that this shift was required because the Al source in [3.22] was contaminated and had a lower X-ray flux than assumed. A repeat of the experiments with a fresh Al surface showed agreement with Fig. 3.12.

We can use Fig. 3.12 to define a sensitivity and a contrast for the resist. The exposure required to give a certain ratio of the dissolution rates, say, $S/S_0 = 10$, can be used to define[2] the sensitivity of a resist (here 300 J cm⁻³). This definition is useful for masks that have a very high contrast, such that the areas under the opaque regions actually develop with the rate S_0. For masks of lower contrast, the dissolution rate ratio $S(E_{inc})/S(T_{abs}E_{inc})$ gives a better characterization of the development process. If we use a certain value (say 10) to define the sensitivity of

---

[2] In resist characterization, one usually defines sensitivity by the required exposure dose so that more sensitive materials are characterized by a lower number.

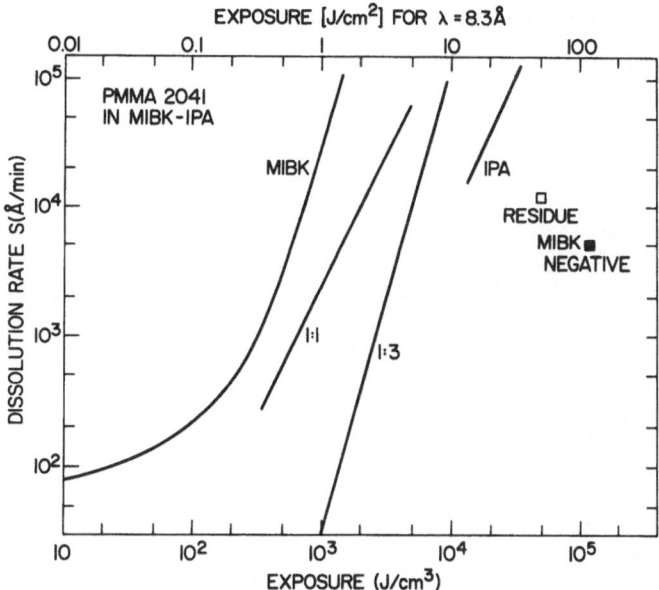

Fig. 3.12. Dissolution rate versus exposure for PMMA developed in methy isobutylketone (MIBK) and isopropyl alcohol (IPA) mixtures. At exposures above 10^5 J cm^{-3} crosslinking begins to dominate and the exposed resist is less soluble than the unexposed

the resist, then the sensitivity would be a function of the contrast of the mask used to test the resist. Figure 3.13 shows some SEM (scanning electron microscope) micrographs of resist patterns obtained for different exposures and mask contrast. For low exposures or low contrast masks, the walls of the resist patterns are sloping. During the time required to remove the exposed resist completely, the boundary between the exposed and unexposed areas in the resist is also partly attacked by the developer and recedes. At the top of the resist this development towards the side occurs for a longer time. The result is a tilted wall with a tilt angle $\beta (\beta = 90°$ for vertical walls) given by:

$$\tan \beta = \frac{S(E_{\text{inc}})}{S(T_{\text{abs}} \cdot E_{\text{inc}})} - 1. \tag{3.14}$$

Equation (3.14) shows how definitions of the sensitivity by the dissolution rate ratio obtained from Fig. 3.12 or by the wall slope as derived from micrographs as in Fig. 3.13 are related. For masks of very low contrast ($T_{\text{abs}} \rightarrow 1$), the slope of the log (Diss. Rate) vs log (Exposure) curve becomes important for the quality of the pattern. The relation

$$\log \frac{S(E_{\text{inc}})}{S(T_{\text{abs}} \cdot E_{\text{inc}})} = \gamma \cdot \log \frac{E_{\text{inc}}}{T_{\text{abs}} \cdot E_{\text{inc}}} \tag{3.15}$$

shows that small ratios between different exposures can become larger ratios in the dissolution rate for $\gamma > 1$. The value γ in (3.15) could be defined as the contrast

MASK CONTRAST: HIGH   MASK CONTRAST: LOW

$E_{inc} (J/cm^2)$

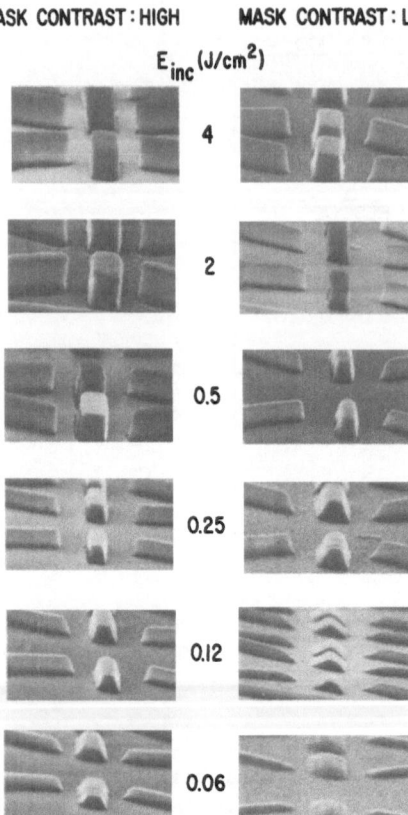

4

2

0.5

0.25

0.12

0.06

$1\mu m$

Fig. 3.13. Resist profiles in PMMA obtained with AlK_α radiation (excitation voltage 8 kV) for different exposure doses. High contrast mask (0.9 μm gold on 6 μm mylar substrate) is used at left, low contrast mask (0.4 μm gold) at right. SEM micrographs with a viewing angle of 80° from normal. Initial resist thickness was 1.5 μm and is equal to the resist thickness after development for the highest exposures

of the resist. It is the analog of the γ value used to characterize the contrast of photographic film. Sensitivity could be defined by the exposure which produces a certain γ value. The slope γ is a very convenient way to characterize a resist, if it is used to produce an image of an object as in X-ray microscopy.

Normalized Remaining Resist. Another way to characterize a resist is to expose it with a test pattern and develop it until the exposed resist is just removed (the unexposed for negative resists). The remaining resist in the unexposed areas (in the exposed areas for negative resist) is normalized to the initial resist thickness and plotted against the exposure [3.65]. For resists like PMMA, for uniform exposure density within the resist, and for patterns with moderate aspect ratios, the resist thickness d decreases linearly with the development time t.

$$d = d_0 - St. \tag{3.16}$$

The normalized remaining resist thickness at the end of the development ($d_{exp} = 0$) is given by

$$d_{norm} = \frac{d_{remain}}{d_0} = 1 - \frac{S(T_{abs} \cdot E_{inc})}{S(E_{inc})}. \tag{3.17}$$

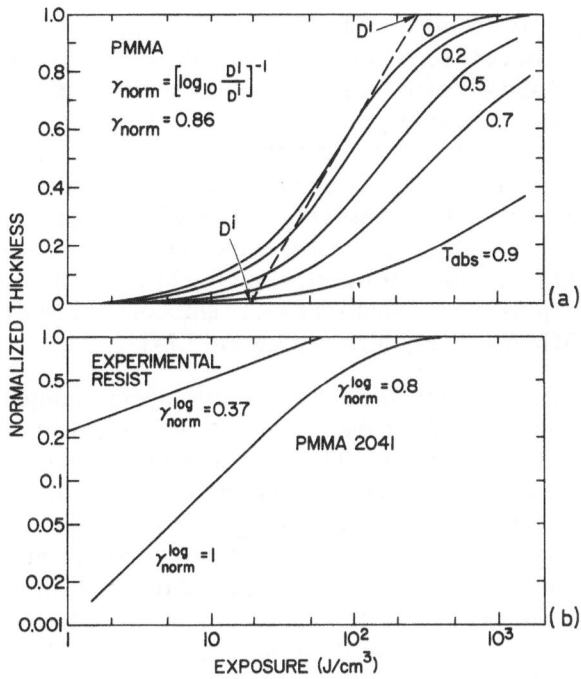

Fig. 3.14a and b. Normalized remaining resist thickness in the unexposed part after completion of the development (complete resist removal in the exposed part) versus exposure with the definition of contrast (a) γ_{norm} or (b) γ_{norm}^{log}. (a) Parameter is the normalized transmission in the opaque part of the mask. (b) Curves are for high contrast masks ($T_{abs} < 0.1$). D^i and D^1 are often used in the literature for the initial and final exposure close E_{inc}^0 and E_{inc}^1.

The normalized remaining resist thickness d_{norm} depends on the contrast ($C = 1/T_{abs}$) of the mask used for the test. Figure 3.14a gives a plot of d_{norm} for different values of T_{abs} derived from the dissolution rate curve $S(E_{inc})$ of PMMA in MIBK in Fig. 3.12. The curve for $T_{abs} = 0$ is often used to define a contrast and a sensitivity for a resist. The sensitivity, for example, can be defined by the value of $d_{norm} = 0.5$ which requires an exposure density of $70\,\mathrm{J\,cm^{-3}}$ in Fig. 3.14a and corresponds to a value of $S = 2 \cdot S_0$ in Fig. 3.12. Contrast is often defined by the exposure range over which d_{norm} changes as

$$\gamma_{norm} = \left(\log_{10} \frac{E_{inc}^1}{E_{inc}^0}\right)^{-1}, \tag{3.18}$$

where E_{inc}^1 and E_{inc}^0 are the exposure values for $d_{norm} = 1$ and $d_{norm} = 0$ obtained by extending the linear part of the curve to these values (Fig. 3.14a). We can plot d_{norm} on a logarithmic scale and define a contrast (γ_{norm}^{log}) as the slope of that curve (Fig. 3.14b). The definitions in Figs. 3.14a and 3.14b give very similar values for the contrast γ_{norm}. The contrast γ_{norm} of a resist, however, is not related in a simple way to the capability of a resist to replicate very low contrast masks. We have to compare curves obtained with masks of different contrast to evaluate this capability of a resist. For a mask with a contrast which is sufficiently low to approximate the dissolution rate curve of Fig. 3.12 by a straight line with slope γ between the exposure values E_{inc} and $T_{abs} \cdot E_{inc}$, the normalized remaining thickness is given by

$$d_{norm} = 1 - T_{abs}^{\gamma}. \tag{3.19}$$

Very low contrast masks ($T_{abs} \to 1$) can only be copied with a value d_{norm} close to 1 if $\gamma \gg 1$. In Fig. 3.12 the highest values of γ are obtained for high exposures, and the highest value of γ between 3 and 4 is suggested from the curves. However, the data points for the highest exposures have large experimental errors, and curves with much larger slopes might still be consistent with the experimental data. A better value for γ than that obtainable from a dissolution rate measurement can be obtained by replicating very low contrast masks ($T_{abs} \to 1$) and by using (3.19) to obtain γ from measured values of d_{norm}. Quantitative measurements of this type have not been performed up to now. However, some results in the X-ray microscopy of low contrast objects suggest that very high values of γ (possibly $\gamma \approx 10$) can be obtained in PMMA for very high exposures [3.74].

Comparison of Different Characterization Methods. For the resist of our example (PMMA), we could derive all characterization methods from a measurement of the dissolution rate as a function of the exposure. The dissolution rate, therefore, gave us a complete description of the resist behavior. The development time is eliminated in a plot of normalized remaining thickness d_{norm}; only the final resist pattern is inspected and used for the evaluation. Required development times have to be provided in addition to the curves of the normalized thickness for a full characterization. The different characterizations of resist contrast and resist sensitivity are based either on the ratio S/S_0 directly, on simple functions of S/S_0 (as d_{norm} and the wall slope $\tan \beta$), or on the slope γ of the dissolution rate curve. All definitions based on S/S_0 are suitable for the characterization of resists for use with masks of very high contrast. In the other extreme where very low contrast masks or objects are replicated in the resist, the slope of the dissolution rate γ gives a better characterization.

In comparing two resists to each other, in many cases a resist which has a higher contrast with one definition has also a higher contrast with another definition and a resist of higher sensitivity has a higher sensitivity independent of which definition is used. However, it is clear from our discussion that this is not true in general. A resist can have a very low value of S_0 and appear as a high contrast and high sensitivity resist if a characterization based on S/S_0 is used, but it may never or only at very high exposures reach a high γ value. Even if the same plot is used, the ratio in contrast and sensitivity obtained may remain a matter of choice. In Fig. 3.14b we have plotted measured values of the normalized thickness for PMMA and for a copolymer of methylmethacrylate and methacrylic acid CP(MMA–MAA) [3.75], which appears to have a higher sensitivity. The gain in sensitivity of the copolymer over PMMA is about a factor of 4 if we choose a value of $d_{norm} = 0.9$ to define the sensitivity. The sensitivity gain is a factor of 20 for $d_{norm} = 0.2$ and 200 for $d_{norm} = 0.1$. Some device fabrication processes may require a value of $d_{norm} > 0.9$; some others can use $d_{norm} = 0.1$. Therefore, all these widespread values for the sensitivities may be meaningful and correct. The curves in Fig. 3.14b are obtained from masks with very high contrast ($T_{abs} = 0.01$). PMMA would become the more sensitive resist, if we would use the value $d_{norm} = 0.9$ for a low contrast mask (e.g., $T_{abs} = 0.5$) to define sensitivities. It is clear from this example that extreme care is necessary when data about contrast and sensitivity from the literature are compared to each other.

We have used an extremely simple development model to compare the different characterization methods for resist. We assumed that there are only two exposure levels in the pattern and that the exposure is uniform throughout the thickness of the resist. We further assumed that this exposure determined the dissolution rate S in a unique way independent of any other parameter, such that each resist surface in contact with the developer recedes with a speed S determined only by the exposure. These assumptions are a good approximation for PMMA for sufficiently thin resists and sufficiently coarse pattern. In general, the development process is much more complicated. Computer simulations of the development have been described; these simulations calculate the shape of the resist pattern as a function of development time and allow a theoretical comparison of different characterization methods, even for more complicated development processes [3.76–80].

For many resists the dissolution rate model has no meaning at all. The development process may not be a surface reaction; a resist may swell, crack, and then float off in small chunks in either the exposed or unexposed parts. Only the final resist pattern is then meaningful to characterize the resist. Negative resists which become insoluble in the exposed areas (due to crosslinking) are examples in which the dissolution rate has often lost its meaning. A crosslinked resist does not dissolve, but if it is not completely crosslinked, a small crosslinked particle might float away due to some soluble material surrounding it. For this and similar reasons, negative resists are usually not described by a dissolution rate curve, but by curves which give the remaining thickness as in Fig. 3.14. In contrast to Fig. 3.14, where the unexposed remaining thickness is plotted, for negative resists the remaining exposed thickness is plotted for a development process which removes the unexposed areas completely.

Resolution

The limit for the resolution of an X-ray resist is the effective range δ of secondary electrons produced in the resist. These secondary electrons are responsible for nearly all chemical changes in the resist [3.8]; each location of an absorption event is therefore surrounded by a small volume of dimensions δ where the resist has been modified. For PMMA this effective range has been measured and found to be equal to the resolution of the resist [3.43, 81]. About the same effective range is to be expected for other organic resists with about the same composition and density. The poorer resolution of these other resists is due to other causes.

For the measurements of the effective range, the resist was overcoated with a thin film of a heavy material which absorbs X-rays strongly and emits electrons into the underlying resist. Erbium was chosen, mainly because it can be removed very easily by dilute acids after the exposure. The effective range of the electrons

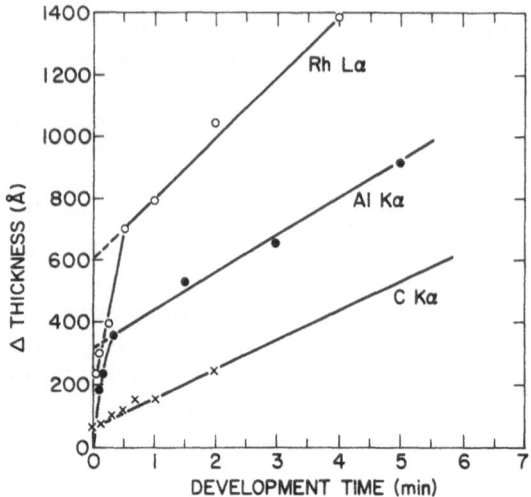

Fig. 3.15. Difference in the resist thickness of exposed and unexposed areas of a resist film, which was overcoated with a thin film of erbium during the exposure, plotted versus development time. The kinks in the curves represent the range of electrons emitted from the erbium film into the resist

emitted into the resist film can be easily recognized in the development curves of Fig. 3.15. The additional exposure by the secondary electrons emitted by the erbium results in a faster development time; the depth at which the transition to the normal development rate due to the X-ray exposure alone occurs represents the range. The measurements showed that the effective range increases linearly with the energy of the incident X-rays ($\delta = 50$, 350, and 650 Å for $E = 0.277$, 1.48, and 2.7 keV). We can equate the ranges measured in Fig. 3.15 with the resolution of the resist, if we assume that the range is primarily dependent on the photon energy of the incident energy and that the differences in the range between secondary electrons generated in erbium or in resist materials can be neglected. PMMA has a better resolution for lower photon energies or longer X-ray wavelengths. At about $\lambda = 50$ Å, the resolution due to the range of secondary electrons is equal to the X-ray wavelength. At still longer wavelengths the resolution is not limited by resist properties but by diffraction effects. The wavelengths region around $\lambda = 50$ Å, therefore, represents the region with the highest possible resolution, at least with PMMA as resist material.

Sensitized X-Ray Resists

Thin organic films have a very small X-ray absorption (Fig. 3.2); a 1 μm thick film of PMMA absorbs less than 10 % of the incident energy of Al K_α radiation with $\lambda = 8.3$ Å. Still lower absorption values are obtained at shorter X-ray wavelengths. The incorporation of heavy atoms into existing resist materials will increase their X-ray absorption and can produce a material of higher sensitivity. It is also possible to sensitize an X-ray resist to a specific X-ray wavelength by using a heavy element with an absorption edge at slightly longer wavelength. Heavy materials can be incorporated chemically into the molecule of the monomer, or they can be dispersed physically in the polymer, for instance in

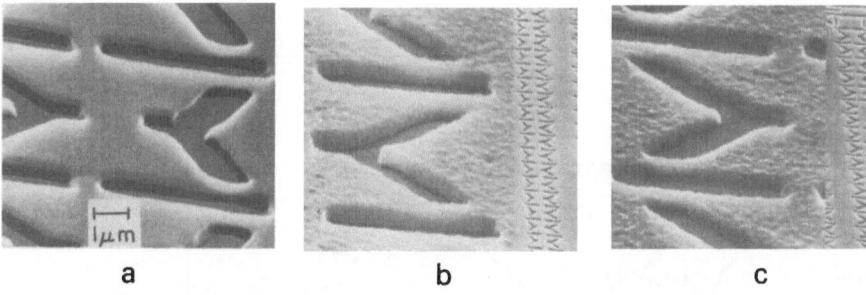

a   b   c

Fig. 3.16a–c. Resist patterns of 1 μm linewidth obtained with Tl doped resist TLP(MM-MA) at exposure levels of (a) 140, (b) 100, and (c) 24 mJ cm^{-2} with AlK_α radiation ($\lambda = 8.3$ Å) developed in a 2.5 mixture of ethoxy ethyl acetate and ethylalcohol with about 2–5% acetic acid

colloidal form. The overcoating with a heavy material as mentioned above can be considered as a form of sensitizing which enhances the sensitivity of the underlying resist. Generally, one might expect that the chemical incorporation will produce a more controllable resist while the dispersion in colloidal form might allow a wider range of dopants to be used.

Most work on sensitized X-ray resists started from known electron-beam resists by modifying these resists to incorporate a heavy material. In several successful cases a gain in the resist sensitivity has been obtained which was larger than the gain in sensitivity expected from the increase in absorption alone; the chemical modifications required for the doping produced an additional increase in the sensitivity. A variety of materials has been used as resist sensitizers, including Ba, Pb [3.82]; Fe [3.83]: Cl [3.37]; Br [3.84]; Tl, Cs [3.85]; and Er, U, Au [3.86]. Figure 3.16 shows resist profiles of a doped resist which is a modified PMMA, doped with Tl. To allow incorporation of the Tl atom into the monomer, the copolymer of methyl-methacrylate (MMA) and methacrylic acid (MAA) was formed [3.85]. The Tl-salt of this copolymer represents the doped resist. Figure 3.16 shows that the gain in sensitivity is much higher than the factor of 3–5 which can be expected from the absorption values alone. In contrast to a developed PMMA pattern, the doped resist develops a grainy structure after long development times. This grainy structure limits the resolution to about 0.1 μm, in agreement with our general expectation that a resist of higher sensitivity has to have a lower resolution. The Tl-doped resist is not in use today because of difficulties in reproducing its properties from one resist batch to another.

The most sensitive X-ray resists which are in practical use today are chlorine-doped negative resists [3.37]. Chlorine has an absorption edge at $\lambda = 4.40$ Å, such that this resist is especially suited for use with PdL_α X-rays of a wavelength $\lambda = 4.37$ Å. The chlorine content of this resist increases the resist absorption by a factor of 10 compared to the value for PMMA. Figure 3.17 gives the normalized resist thickness curve of Poly (2.3-dichloro-1-propyl acrylate) (DClPA). For $d_{norm} = 0.5$, an absorbed dose of 7.8 J cm^{-3} is required compared to PMMA which requires 70 J cm^{-3}. The differences in sensitivity become more drastic if

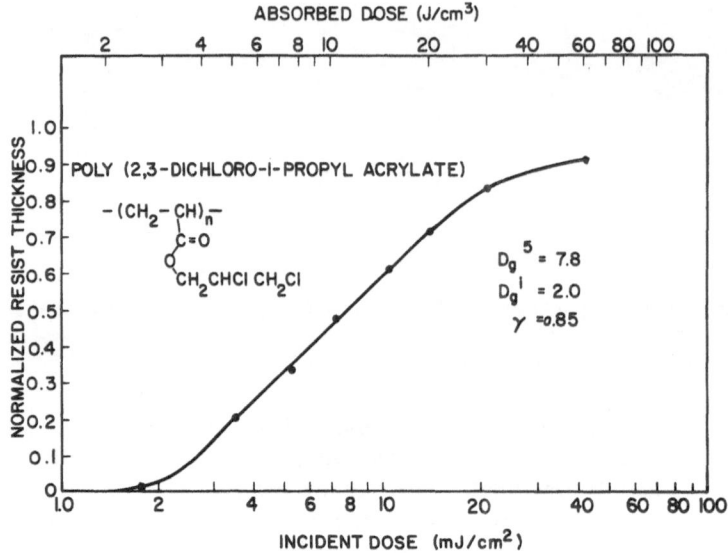

Fig. 3.17. Normalized remaining resist thickness for Cl-doped resist DClPA versus exposure. Top scale: absorbed X-ray dose, bottom scale: incident dose for $\lambda = 4.37$ Å. $D_g^{1.0}$ and D_g^i correspond to E_{inc}^1 and E_{inc}^0 in Eq. (3.18) (courtesy of *Taylor* [3.37])

we compare the required incident doses. The absorption of PMMA for $\lambda = 4.37$ Å is $\alpha = 1400\,\mathrm{cm}^{-1}$, while we have $\alpha = 1400\,\mathrm{cm}^{-1}$ for the chlorine resist [3.37]. Therefore, the $70\,\mathrm{J\,cm}^{-3}$ required for PMMA corresponds to an incident dose of $0.5\,\mathrm{J\,cm}^{-2}$ for $\lambda = 4.37$ Å compared to $0.0078\,\mathrm{J\,cm}^{-2}$ for the chlorine-doped resist.

Resist Summary

Table 3.2 summarizes the properties of polymer films which have been used as X-ray resists. Sensitivities are given as the incident required exposure E_{inc} and as the absorbed power in the resist, which is given by αE_{inc}, where α is the linear absorption coefficient of the resist materials. The linear absorption coefficient is calculated from the mass absorption coefficients of the elements [3.12–14] and the chemical composition of the resist. For those resists for which data on the resolution δ are published, these data are included and the value $\bar{n}$, the number of photons absorbed in a cube with dimensions δ^3 at the required exposure, is also given. The exposure of $500\,\mathrm{J\,cm}^{-3}$ for PMMA corresponds to a value $\bar{n} = 1.4$ for $\lambda = 44.8$ Å and an assumed resolution $\delta = 50$ Å. While micrographs with a resolution close to 50 Å have been obtained with considerably higher exposure levels (about $10^4\,\mathrm{J\,cm}^{-3}$) [3.52], we can still conclude that PMMA has a performance which is very close to that of an ideal resist around $\lambda = 50$ Å and that higher sensitivities, measured by the absorbed power, cannot be expected from any other resist with the same resolution. However, because a 50 Å thick

Table 3.2 Properties of several X-ray resists. Unless noted, the required incident dose is for $\lambda = 8.3$ Å. The absorbed dose is obtained by multiplying the incident dose with the linear absorption coefficient α of the resist material

Resist	Characterization	Required incident dose [J cm⁻²]	Required absorbed dose [J cm⁻³]	Resolution δ	$\bar{n}$ absorbed photons/δ^3
Polymethyl methacrylate (PMMA), positive [3.8, 9, 20, 70, 78]	$\gamma = 2$	0.33	330	350 Å ($\lambda = 8.3$ Å)	500
	$\gamma = 3$	0.5	500	50 Å ($\lambda = 44.8$ Å)	1.4
	$\gamma = 3$	3.5 ($\lambda = 4.37$ Å)	500		
	$d_{norm} = 0.5$	0.07	70		
	$d_{norm} = 0.9$	0.2	200		
CoP(MMA-MAA) modified PMMA to allow doping, positive [3.50, 75]	$\gamma = 1$	0.05	50		
	$d_{norm} = 0.5$	0.01	10		
	$d_{norm} = 0.9$	0.05	50	< 500 Å	< 260
TIP(MMA-MAA), positive [3.85]	$\tan\beta = 5$	0.024	80	1000 Å	340
Poly(butene-1) sulfone, positive [3.83,87]	$d_{norm} = 0.9$	0.01	14		
Poly(vinyl ferrocene)	$d_{norm} = 0.5$	0.12	250		
PVFc, negative [3.83]	$\gamma_{norm} = 1.5$				
Kodak Micro-Negative Resist, KMNR, negative [3.88, 89]	$d_{norm} = 0.5$	0.06	60		
Terpolymer of MMA, ethylacrylate and glycidylmethacrylate CER, negative [3.83]	$\gamma_{norm} = 0.78$	0.025	25		
Barium lead acrylate [3.82]	$d_{norm} = 0.5$	0.025	60		
	$\gamma_{norm} = 1.2$				
Polybutadiene, PB, negative [3.83, 87]	$d_{norm} = 0.5$	0.04	20		
	$\gamma_{norm} = 1$				
Poly(diallyl orthophthalate), PDOP, negative [3.83]	$d_{norm} = 0.5$	0.014	14		
	$\gamma_{norm} = 0.78$				
Poly(glycidyl-methacrylate-ethyl acrylate), P(GMA-EA), negative [3.83]	$d_{norm} = 0.5$	0.005	5		
	$\gamma_{norm} = 1.07$				
Poly(dichloro propyl acrylate), DCIPA, negative [3.37]	$d_{norm} = 0.5$	0.0078 for $\lambda = 4.37$ Å	10	0.5μm	5000
	$\gamma_{norm} = 0.85$				
Epoxidized polybutadiene, negative [3.83, 87, 90]	$d_{norm} = 0.5$	0.0015	1	1 μm	4000
	$\gamma_{norm} = 0.56$				

layer of PMMA absorbs only about 10^{-3} of the incident power, this statement is not true if sensitivity is measured by the incident power. An ideal detector could absorb all of the incident photons and would therefore require only 1/1000 of the exposure for the same signal/noise requirement. The high sensitivity, medium resolution resists in Table 3.2 require many more absorbed photons per resolution element than PMMA. A further improvement in the sensitivity appears to be much more easily achievable for a high sensitivity resist like DCPA than for a low sensitivity, high resolution resist like PMMA.

3.7 Alignment

Most electronic circuits require several fabrication steps for which different patterns have to be produced on the device wafer at different times. These patterns have to be overlayed with great accuracy. As a rule of thumb, one requires an overlay accuracy of about 1/5 of the smallest linewidth. Existing optical alignment techniques[3] [3.91–95] can overlay a mask over a pattern accurately to about 0.25 μm, thus allowing fabrication of devices with linewidths larger than about 1 μm. The optical alignment techniques can be adapted to X-ray lithography, provided that either the mask or the wafer is transparent to light or has at least alignment windows that are transparent. Optical alignment of X-ray masks has been used successfully to fabricate silicon devices with linewidths of 2.5 μm [3.19, 38, 89].

Modified alignment systems will be required if one tries to fabricate multilevel devices with linewidth less than 1 μm. Several proposals as to how the required alignment might be obtained have been published [3.9, 88, 96–98, 103], and experimental systems have obtained a positioning accuracy around 0.1 μm [3.99, 100].

For experimental work in the laboratory with the goal of fabricating circuits with extremely small geometries which might have unique properties, the required alignment accuracy represents no fundamental problem. However, for commercial applications, where the driving force for smaller device geometries is lower cost, problems can be anticipated. Lower cost has in the past been obtained by maintaining or even enlarging the size of a wafer when proceeding to smaller geometries. The larger number of circuits per wafer reduced the cost per circuit drastically because the processing cost per wafer did not change very much and because wafer and masks were mechanically stable enough to allow overlay of the full wafer with the required accuracy.

For the fabrication of circuits with 0.5 μm smallest linewidth, an alignment accuracy of about 0.1 μm has to be expected. Wafer and mask with a 10 cm diameter have to be stable to 1 part in 10^6 to allow the required alignment for the full wafer. A stability of $1:10^6$ is already nontrivial to maintain for elements in a well-protected environment. Device wafers go through processing steps between

---

[3] Kasper 2001 A alignment system, Kasper Instruments, Inc., Mountain View, CA.

subsequent lithographic exposures which add or remove material from these wafers. Very often, as in the case of silicon devices, these processing steps include high temperature processes around 1000 °C which will distort the wafers considerably. Exact numbers about these distortions are hard to find; a general consensus at the present time seems to be that the dimensions of silicon wafers throughout the processing steps are stable to about 0.25 μm. No data exist at the present time which allow us to predict how far this stability might be improved with modified processing steps and circuit designs which try to minimize the sources of the distortion.

Mask stability may also become a problem for small linewidth circuits. We think, however, that these problems will not be as severe as those caused by wafer distortions. Masks can be kept in a stable environment and there are indications that a heavy exposure with X-rays accumulated over many mask replications does not change the dimensions of a mask [3.50]. Intuitively, many people, especially those who have never handled X-ray masks, have little confidence in the mechanical stability of a very thin membrane. It may be worth-while to note that the stability of an isotropic thin membrane of uniform thickness, stretched over a rigid frame, is completely determined by the stability of the rigid frame, as long as the tension in the membrane is high enough to keep it taut under all conditions. An increase in temperature, for instance, may relax some of the stress but without changing the dimension more than in a rigid plate of the support material of the frame.

The stability problem is a basic problem for any type of lithography with fixed masks. Replication with slight magnification (as copying with an adjusted mask wafer gap in a slightly diverging beam, Fig. 3.1) may compensate for some part of the distortion. The random, pattern-related distortions will become one limit for further miniaturization using X-ray lithography. Step and repeat exposures of small areas of a wafer may overcome some of the problems, but will make the exposure system more complex. It is obvious that an electron beam system which exposes wafers point by point does not suffer from the stability problem because it can measure the wafer distortions and adjust the beam position locally.

3.7.1 Possible Alignment Configurations

For the following, we shall assume that the stability problem is solved and discuss some of the possible alignment methods for X-ray lithography. Only optical alignment has actually been used; the other methods have up to now not been tested in practice. Figure 3.18 gives a summary over possible geometries between mask and wafers and also lists methods to detect alignment or misalignment. The configurations A, B, C, and D are the four combinations if either mask or wafer or both are transparent for the radiation used for the alignment. E and F represent the case where a secondary effect produced by the incident radiation is used for the detection of alignment, and G symbolizes all cases without incident radiation, where a field generated on the mask (or wafer)

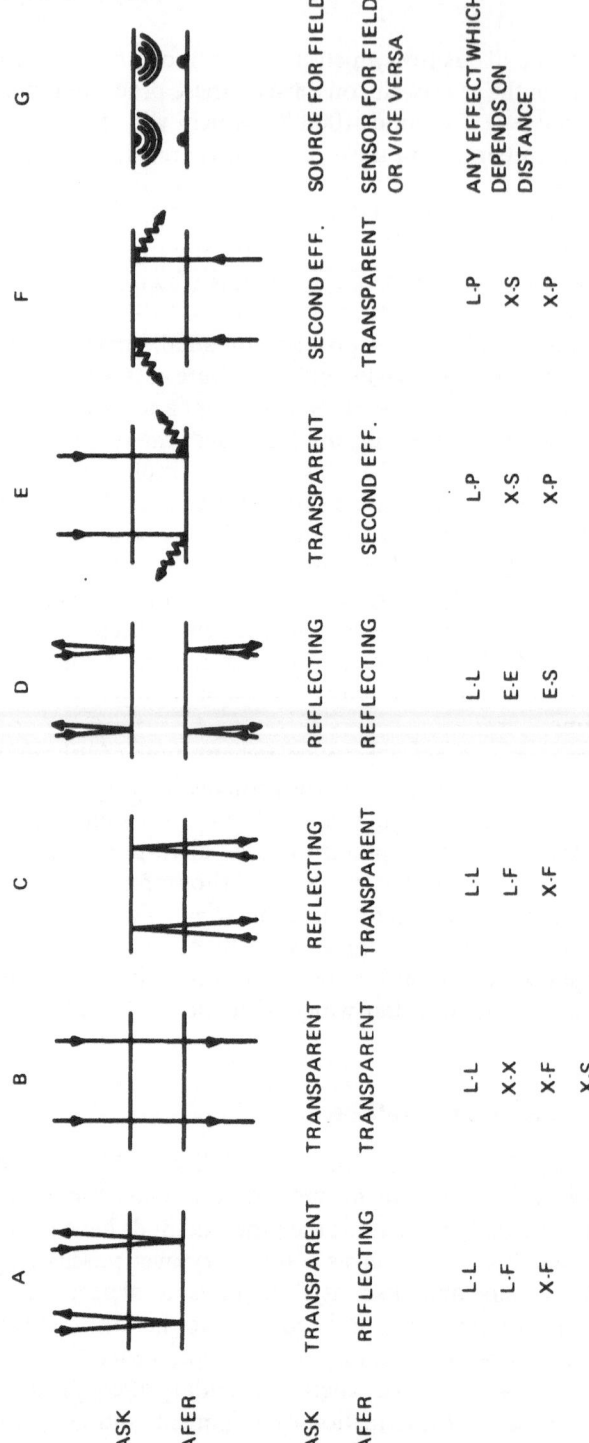

	A	B	C	D	E	F	G
MASK	TRANSPARENT	TRANSPARENT	REFLECTING	REFLECTING	TRANSPARENT	SECOND EFF.	SOURCE FOR FIELD
WAFER	REFLECTING	TRANSPARENT	TRANSPARENT	REFLECTING	SECOND EFF.	TRANSPARENT	SENSOR FOR FIELD OR VICE VERSA
	L-L	L-L	L-L	L-L	L-P	L-P	ANY EFFECT WHICH
	L-F	X-X	L-F	E-E	X-S	X-S	DEPENDS ON
	X-F	X-F	X-F	E-S	X-P	X-P	DISTANCE
		X-S					

Fig. 3.18a–g. Possible geometries for mask wafer alignment and combinations for illumination and detection which are compatible with these geometries. L = light, X = X-rays, E = electron beams, F = fluorescence, S = any secondary effect, P = photoconductivity

is sensed at the wafer (or mask). For each case we have also listed examples for the type of radiation or effects which might be used for illumination and for detection. (L = light, X = X-rays, E = electron beams, F = fluorescence, S = secondary electrons, P = photoconductivity.)

Optical Alignment

All alignment machines in practical use today use light and correspond to A in Fig. 3.18 (L–L). An alignment mark on the wafer is aligned in respect to a corresponding mark on the transparent mask. In X-ray lithography when often a considerable gap between mask and wafer is used, it is convenient to use a microscope which can bring both planes into focus simultaneously. Possible designs can use either different wavelengths or different polarizations for the mask and wafer image or could compensate for the different distance by rotating a transparent plate with two different thicknesses through the viewing beam [3.101].

Optical measurement instruments can measure distances between marks to an accuracy better than 0.1 µm [3.102]. The reason that alignment machines in device fabrication have obtained only about 0.25 µm accuracy as a best value may be partly due to the stability of the silicon wafers. The main difficulty is, however, that the alignment marks on the wafer in a manufacturing system are exposed to all the wafer processing steps which may alter them. In addition, resist of variable thickness and profile covers the alignment marks during the alignment. Well-protected alignment marks not covered by resist during the alignment should allow alignment better than 0.1 µm with optical systems. The main advantage of configuration D (Fig. 3.18), where alignment marks are on the back of the wafer, is the easier protection of alignment marks during the device processing steps.

X-Ray Alignment

Alignment with X-rays promises higher resolution than does alignment with light. In addition, more reproducible results should be obtained because the refractive index of all materials is very close to one so that changes in the topography will not deflect the radiation. Because X-rays are practically not reflected by any material, transmitted X-rays have to be used for the alignment (configuration B), or a secondary effect (configuration E) has to be used to produce an alignment signal.

For X-ray alignment in transmission, the device wafer has either to be thinned in the alignment windows [3.9, 88] or a short wavelength for which a thick wafer has sufficient transmission has to be used [3.97, 98]. The use of a thinned wafer window gives the strongest alignment signals and therefore the best signal/noise ratio and the shortest alignment time. Alignment with harder, more penetrating radiation, has a poorer signal to noise ratio due to the lower contrast of the alignment mark. *McCoy* and *Sullivan* estimated that reasonable alignment times can still be obtained with hard X-rays ($\lambda = 1.2$ Å) and 200 μm thick silicon device wafers [3.97, 98].

Alignment which uses secondary effects produced by a pattern on the wafer which matches the absorber pattern on the mask allows arbitrary thick device wafers and soft X-rays for the illumination. X-ray fluorescence [3.96], X-ray photoemission [3.104], or the X-ray induced current in a patterned solid state device like an MOS structure [3.105], have been proposed as the detection method.

An alignment system will generally align only one mask wafer combination at one time. A conventional X-ray source on the other hand can expose many wafers simultaneously. This mismatch results either in a drastic decrease in throughput (if only one wafer at a time is exposed) or a drastic increase in cost (if an alignment system is built for each exposure position) for a system with alignment capability. A separate alignment system which produces a batch of aligned mask wafer combinations for later exposure could eliminate that disadvantage. At the present time there exists no practical experience as to how well a system which has to lock the aligned mask wafer combination for the transfer from the alignment to the exposure station would work.

A big advantage of X-ray lithography is its large depth of focus which allows large gaps between mask and wafer. A variation in the gap, however, presents a change in the magnification, if the exposure is performed with a diverging X-ray beam as in Fig. 3.1. For such systems, gap control becomes important if different exposures are registered to each other. For a mask wafer combination which uses a cone of radiation within a half angle of 0.1 rad, the gap has to be controlled within 1 μm for a control of the geometric distortion to 0.1 μm. Highly collimated X-ray sources like synchrotron radiation sources relax the requirements on gap control drastically. Because lithographic systems using synchrotron radiation in general will expose only one wafer at a time (with drastically shorter exposure times), the addition of an alignment system does not decrease throughout. Synchrotron radiation sources show their advantages over conventional sources most drastically for exposure with alignment.

3.7.2 Self-Aligning Methods

A method to produce several device fabrication levels with one exposure only, thus eliminating the need for alignment between the levels, was first proposed by *Henderson* and *Pease* [3.106] for electron beam lithography. The basic idea is to distinguish the different fabrication levels by different exposure densities in one single exposure. In a first development step, the resist is opened for the first processing steps only in those parts of the pattern which obtained the highest exposure; a subsequent development opens up the second level for processing and so on. It is clear that this method will require tighter tolerances on exposure and development and also poses some restriction for the subsequent processing (protection of the first levels during processing of later levels; high temperature

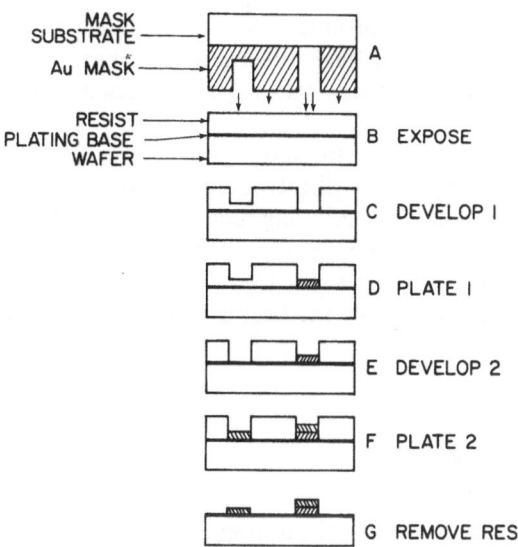

Fig. 3.19. Processing steps to fabricate two aligned metallizations from a single exposure with a two-level mask

processing only during the last level, otherwise the resist which carries the information for the later levels might be destroyed). The tighter tolerances represent a severe limitation for the use of the method in electron beam lithography where tight tolerances are already required to control electron scattering and proximity effects [3.107–109]. X-ray lithography has very much looser tolerances on exposure and development, so the method looks much more promising here. In X-ray lithography the different fabrication levels would be represented in the mask by different thicknesses of the absorber patterns giving different transmission values and consequently different exposure levels in the mask (Fig. 3.19). The method has been used successfully for the fabrication of a two-level system with X-ray lithography; the example chosen was the different metallization required in the sensor area of a bubble memory device compared to the rest of the memory [3.81].

The method has also been proposed for the fabrication of silicon devices [3.110]. High temperature processes could be tolerated if intermediate steps are introduced which, for instance, first deposit different temperature resistant materials in the areas defined by different development times and then use specific etchants for these materials to open selective areas for further processing when required.

An interesting method for eliminating wafer distortion is the fabrication of a multilevel mask on the back of the wafer [3.111]. The mask could contain several "grey" levels for one wavelength or could be made out of different X-ray absorbers which selectively absorb different X-ray wavelengths. Patterns could be selectively transferred from the back to the front of the wafer by proper selection of X-ray wavelength and intensity for this transfer. A change in the dimensions of the wafer will affect the pattern on the back in the same way as that on the front so that perfect alignment is maintained. All self-aligned methods require that a multilevel mask be produced first. Alignment will in general be necessary for the overlay of the different levels of this mask. High resolution alignment capabilities for an *e*-beam system to produce such a mask do exist [3.112–114].

A lot of detailed work is necessary to adapt circuit design and fabrication to the use of self-aligning methods. The largest disadvantage is some loss of freedom in the choice of processing steps. Self-aligning methods will probably not be able to replace conventional alignment completely but may be found very useful to reduce the number of required alignment steps. Those levels which require the most critical alignment are probably the best candidates for a self-aligned exposure.

3.8 Subsequent Processing Steps

The developed resist pattern is the end product of the lithography. However, this resist pattern is only the starting point for a series of processing steps to define the final device. The resist pattern has to be able to withstand all these processing steps and must also not cause any negative effects on the yield and quality of these steps. The task of making the lithographically obtained resist pattern compatible with all the subsequent processing steps is time consuming and often frustrating. A decision to change some parameter or process in the manufacturing line often changes the requirements for the lithography, while a small change in the lithography (as changing from one batch of resist to another) can generate unexpected problems at a later step. It is, therefore, necessary to know as much as possible about these processing steps to select the best parameters for the lithography. Unfortunately, the basic understanding of many processes is still lacking and often the detailed parameters of a working process are considered proprietary.

We will give a short overview of the many processing steps and discuss their compatibility with X-ray lithography. A more detailed discussion can be found in [3.115], which is a collection of review articles [3.116–120] with several extensive bibliographies. A comparison of several processing techniques for high resolution X-ray lithography is given in [3.121].

Wet chemical etching has been used by artists for 300 years and has also dominated semiconductor fabrication in the past [3.116, 117]. Because in most cases the etching solution etches isotropically, linewidth control is difficult as soon as the linewidth and the thickness of the material to be etched are of the same order. For high resolution work, wet etching is therefore more and more being replaced by other processes. It can still be used for the removal of very thin films, like the plating base of a finished X-ray mask, or for the removal of a thin film which might serve as the protective mask for another process which gives better wall definition. Even in the etching of very thin films, problems can arise. Let us assume we want to remove a 400 Å thick Ni–Fe plating base from the open areas of an otherwise finished Au X-ray mask to make this mask more transparent for ultrasoft X-rays (e.g., $\lambda = 44.8$ Å). We have adjusted the concentration of the etchant ($FeCl_3$) to obtain a well controllable etching time of around 1 min for this film. However, the Ni–Fe film which is under the opaque, gold plated areas of the mask forms a galvanic cell with the gold. The etching rate at the boundary increases about 2 orders of magnitude to an uncontrollable high value resulting in the destruction of the mask pattern [3.117].

Chemical plasma etching or reactive plasma etching [3.119, 122] is very similar to wet chemical etching. The difference is that the liquid etch solution is replaced by a reactive gas, which is usually made reactive by a gas discharge surrounding the etch chamber. Halogens are the most widely used gases for semiconductor manufacture. They are often generated in the plasma from halocarbons. The etch rate of most resists is lower than that of Si or SiO_2 and resists can therefore be used as masks for the etching process. The etch rate for resists is drastically increased if oxygen is present in the plasma [3.119]. Plasma etching with oxygen to remove a resist film is called *ashing*.

Sputter etching and *ion milling* [3.120, 123, 124] are processes in which accelerated ions bombard the surface to be etched and mechanically remove material. If chemical reactions are also used for the removal by adding a reactive gas to the usually used noble gas like Ar, the method is called *reactive ion etching* [3.117]. In sputter etching, the workpiece is mounted on one electrode of the discharge chamber and ions are accelerated towards it. In ion milling, the ions are generated in an extra ion source and bombard the sample which is located in a field-free high vacuum chamber. The directionality of the ion beam produces better defined wall profiles than do the chemical etch processes. The dependence of the etch rate on the angle of incidence determines the slope of the etched walls. For most conditions the highest etch rate occurs not for normal incidence but at some other angle. Walls with that slope, therefore, etch faster and initial small facets at that angle will grow in an ion milling system and produce tilted walls (Fig. 3.20). Etch rates of resists are not drastically lower than the etch rate of a metal like gold. For good linewidth control, relatively thick resist films with vertical walls are required. Due to the higher etch rate at oblique angles the resist surface becomes very rough during the milling (Fig. 3.20); in addition the ion bombardment makes many resists insoluble and it is often difficult to remove the resist after the process. The forming of facets and roughening can be reduced by tilting and rotating the sample during ion milling. In a sputter etching system, where the ions are not as well collimated as in an ion milling system, this procedure is not required.

The *additive lift-off* process [3.127] is a process in which material is added by evaporation in the areas not protected by the resist. The unwanted evaporated material on top of the resist is afterwards lifted off when the resist is stripped. Resist walls with an undercut (Fig. 3.21) are required for this process in order to leave openings in the evaporated film which give the resist solvent access to the resist layer. Resist walls with an undercut can easily be produced by electron

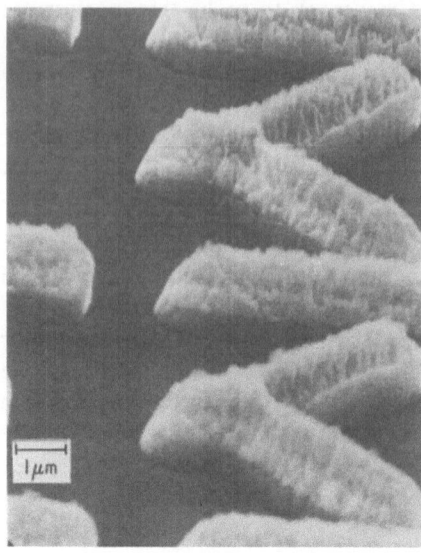

Fig. 3.20. Ni-Fe pattern of 1 μm linewidth ion milled with a mask of X-ray resists demonstrating forming of facets, roughening of the resist film [3.126] and partial redeposition [3.125] of the material at the sides of the pattern (at right side of photo)

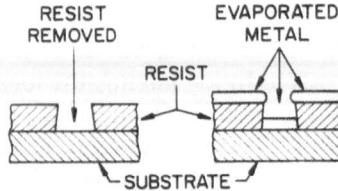

Fig. 3.21. Schematic representation of the lift-off process (courtesy of *Hatzakis* [3.108])

beam exposure due to the scattering of the electrons [3.108, 127]. In X-ray lithography, resist walls with undercut can be produced by a double layer of resist, one resist which develops very slowly at the top and one which develops much faster at the bottom. Figure 3.22 is an example of the wall profile obtainable with such a double layer of resist [3.22].

Additive electroplating [3.22, 128] allows us to add metals in the areas not protected by the resist. The plated material produces an exact replica of the resist walls and the plated material can be deposited up to a thickness equal to the resist thickness. The finest linewidths and the highest aspect ratios in an X-ray mask have been obtained by electroplating. In 2 μm thick resist, 1000 Å wide lines of gold have been plated to the same thickness as wider lines which exist in the same pattern [3.43]. The limit in the linewidth which can be obtained is not known but is probably below 500 Å. While plating gives the highest pattern quality of all processing steps, it has some limitations. It is limited to the more noble metals and their alloys. The surface to be plated must be conductive, which requires the deposition of a thin plating base (e.g., 200 Å Au) before resist is applied to the wafer. The removal of this plating base from the unplated area by

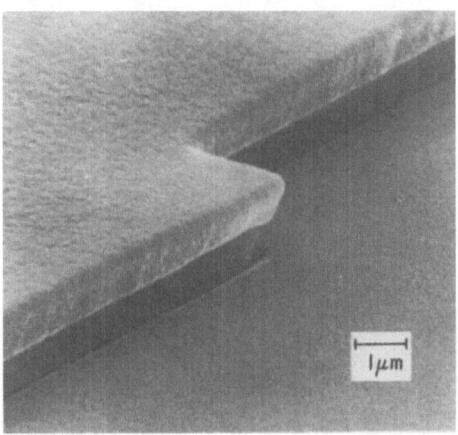

Fig. 3.22. Resist profile with undercut suitable for the lift-off process obtained by X-ray exposure with a double layer of resist. Total thickness of resist is 1.5 μm

any of the dry etching steps is no serious problem but requires an extra step. A thin layer of resist residue left on the substrate after development can be a problem for the subsequent plating as well as for lift-off and chemical etching. It is difficult to recognize such a resist residue, especially at the bottom of very fine lines. Ashing has been used as an extra step to make sure that any resist residue is removed [3.22].

Plating solutions are commercially available[4]; their composition is very complex and proprietary. Such a solution may attack some resists, and traces of the dissolved resist may contaminate the plating bath and change its characteristics. The only definitive test of a bath is the plating result. A monitoring system, which monitors the quality of the bath before and during the plating to allow corrective action before something goes wrong, would be very desirable but does not exist at the present time.

Thermal diffusion processes in semiconductor fabrication are carried out at a temperature which is too high for the resist film. Therefore, the resist pattern has to be replaced by a high temperature mask for this process and the fabrication of this secondary mask uses one of the processes mentioned above (e.g., etching of SiO_2). *Ion implantation* [3.129] has partly replaced the thermal diffusion process in recent years and has been used for semiconductor fabrication and integrated optics. Resist can be used as a mask for ion implantation; therefore, ion implantation can be a primary processing step for lithography. No work on the use of ion implantation in connection with X-ray lithography has been published. Table 3.3 gives a comparison of the most widely used processing steps published first in a review paper by *Romankiw* [3.117].

Very often it can be advantageous to use a combination of processing steps instead of a single processing step [3.130]. An example is the X-ray mask fabrication sequence at Bell Laboratories [3.38]. In this sequence the resist is

---

[4] *Sel Rex Corp.*, Nutley, NJ.

Table 3.3 Comparison of the effect of several process variables on the ability to generate metal pattern by different patterning techniques

	Wet chemical	Dry plasma	Sputter-etching & ion milling	Additive lift-off	Additive electrodeposition
1. Deposition process	Affects etch rate & edge smoothness	Affects etch rate	Affects etch rate	Limited to temp. <80°	Affects thickness
2. Resist adhesion	Must be excellent	Not critical	Not critical	Not critical	Must be excellent
3. Resist edge profile	Not critical	Critical	Critical, but gives some flexibility for pattern profile	Undercut required	Vertical preferred
4. Resist thickness	Not critical above 0.5 μm	Function of relative etch rates	Function of the relative etch rates of the film and of the resist	≥1.5 times the pattern	≥thickness of the pattern
5. Etched line width	Over-etching	Slight narrowing	Can vary from wider than to narrower than resist	Slightly narrower than slot in resist	Same as exposed slot in resist
6. Over-etching	Serious problem; catastrophic in some cases	Great leeway as long as mask holds up	Not serious if resist holds up	If evaporation too thick cannot lift off	Overplating will form a mushroom
7. Etched edge	Angle control difficult (except in some duplex films)	Reported as smooth; angle easy to control near 45°	Depends on the mask & grain size. Can be controlled anywhere from 30 to 90°	Relatively smooth; trapezoidal ~60 to 80°	Exact replica of the resist. Usually perpendicular or inverted trapezoid
8. Selectivity	Good between oxides and metals, poor between metals	Controllable	Very poor; over-etching easy	Lift-off solution should not attack the metal	Plating is very selective but limited to small number of metals; bath may attack resist
9. Resist attack	Nonuniform in strong acids, alkalis	Uniform erosion	If not properly cooled, resist will break down and carbonize		
10. Micron & submicron dimensions	Difficult	Easy	Easy in thin films	Reasonably easy; better than for wet chemical	Very easy, as good as resist exposure
11. Resist removal	Separate operation more difficult in negative working resists	Easy, usually in situ	Usually very difficult	Is part of the lift-off operation, often requires ultrasonic field assistance	A separate operation, also have to remove the plating base

(courtesy of Romankiw [3.117])

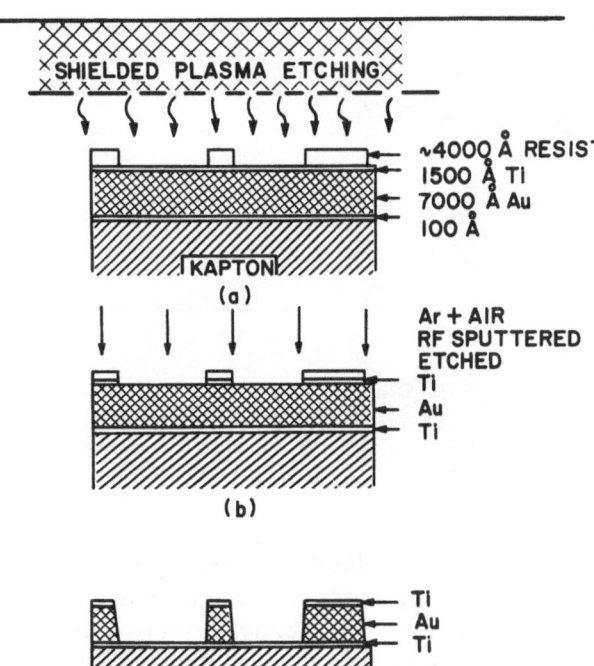

Fig. 3.23. Processing sequence for the fabrication of the gold pattern in an X-ray mask using an intermediate Ti mask for the sputter etching of the Au (courtesy of *Maydan* [3.38])

used as a mask to plasma etch a very thin film of Ti (Fig. 3.23) and then the pattern in the Ti is used to mask the thick Au layer during sputter etching in the presence of oxygen. The advantages of this sequence versus the direct etching of gold with a resist mask are, first, that the requirements for the quality (wall slope) and resist thickness can be very much relaxed, because only a short etching time is required to remove the thin Ti film and during this short time the resist is practically not changed. As a consequence, the exposure time for the resist exposure can be reduced. In the second sputter etch step, a very low etch rate of the thin Ti film is achieved by admitting O_2 in the chamber (probable formation of etch resistant oxides). Further, this low etch rate allows the use of the very thin Ti film and reduces tilting of the Au walls due to faceting and lateral shrinkage of the Ti mask [3.124]. Figure 3.24 shows an SEM micrograph of the gold pattern in an X-ray mask produced with the described processing sequence.

3.9 X-Ray Lithographic Systems

The design of a complete X-ray lithographic system is a compromise between many parameters, and, depending on the weight which one gives to the different parameters, quite different systems are obtained. One can optimize a system with

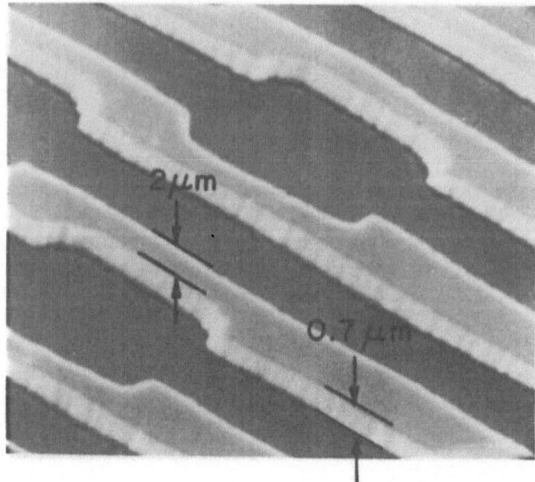

X-Ray mask,
metalization level

Fig. 3.24. Gold pattern in an X-ray mask fabricated according to Fig. 3.23 (courtesy of *Maydan* [3.38])

respect to exposure time [3.41, 131], resolution, or throughput but this optimization is in practice constrained by additional requirements like compatibility with subsequent processing steps, availability of sources, windows, substrate materials resists, and many more. An alignment requirement plays a decisive role in the design of the system.

3.9.1 Parallel Exposure of Many Wafers

A system without the need for accurate alignment can be as simple as that given in Fig. 3.1. Many wafers can be exposed simultaneously with the mask wafer combinations usually mounted on a spherical dome so that nearly all the radiation emitted from the source is used. For extremely high resolution, large distances between source and mask are chosen to reduce penumbral blurring [Fig. 3.1 and (3.1)]. The exposure time increases quadratically with the source wafer distance, but, because the number of wafers which can be exposed during one exposure increases in the same way, the throughput is independent of this distance. The system of Fig. 3.1 can be easily adapted for various requirements, especially if the exposure takes place in a vacuum without any window between source and wafer[5]. Different wavelengths between 4 and 50 Å can easily be obtained by changing only the target material and the accelerating voltage of the gun. With a source of 400 W input power, the throughput of this system is about 40 wafers of 2.5 cm diameter per day (24 h) using PMMA as resist.

Considerably higher throughputs are possible by using resists of higher sensitivity and rotating anode sources of higher power. Table 3.4 summarizes

---

[5] For operation without a window it is important to prevent stray electrons from the gun from being accelerated towards the mask. This can be achieved by keeping the mask and wafer supporting dome at cathode potential.

Table 3.4 Throughput data for X-ray lithographic systems determined by the X-ray flux

System	Throughput
Electron gun with 400 W input power into water cooled target $\lambda = 8.3\,\text{Å}$ mask substrate with 50 % transmission, resist PMMA, exposed for vertical walls	40 wafers (2.5 cm diameter) per day

System change	Factor of increase in throughput
Rotating anode gun	20
Synchroton radiation from storage ring	10^5
Improved resist	100 for about 1 μm resolution

throughput data for systems which use the full emission cone of the source. The throughput is radiation limited for the low power gun and a low sensitivity resist like PMMA; however, throughput is no serious problem for systems with parallel exposure, only moderate easily achievable improvements in resist sensitivity or source intensity are necessary to meet all requirements. Alignment could be provided without any change in the system by an aligner outside the system and a mask wafer lock (see Sect. 3.7).

3.9.2 Single Wafer Exposure

Figure 3.25 shows an X-ray exposure system with alignment built at the Bell Telephone Laboratories [3.38]. A very similar system is used at General Instrument [3.132].

In this system the mask wafer combination is aligned under an optical microscope (dual focus system) to bring both mask and wafer into focus and then shifted under the X-ray source for exposure. Parameters of the system are summarized in Table 3.5. The short X-ray wavelength allows the use of a relatively thick vacuum window so that exposure can take place at atmospheric pressure in helium. With PMMA as resist, the system would require an exposure time of about 1000 min for vertical resist walls and give a throughput of 1 wafer/day. An increase in throughput is obtained by using a chlorine-doped resist of higher absorption and sensitivity (Sect. 3.6) and by selecting subsequent processing steps which do not require vertical walls in thick resist (Sect. 3.8). An exposure time of minutes is required, which makes this system very useful to test the capabilities of X-ray lithography for the fabrication of multilevel semiconductor devices. The large distance of 50 cm between source and wafer was chosen to reduce the geometrical distortion (run-off, see Fig. 3.1) to acceptable levels. The resolution of the system is limited by the resist to linewidths greater than

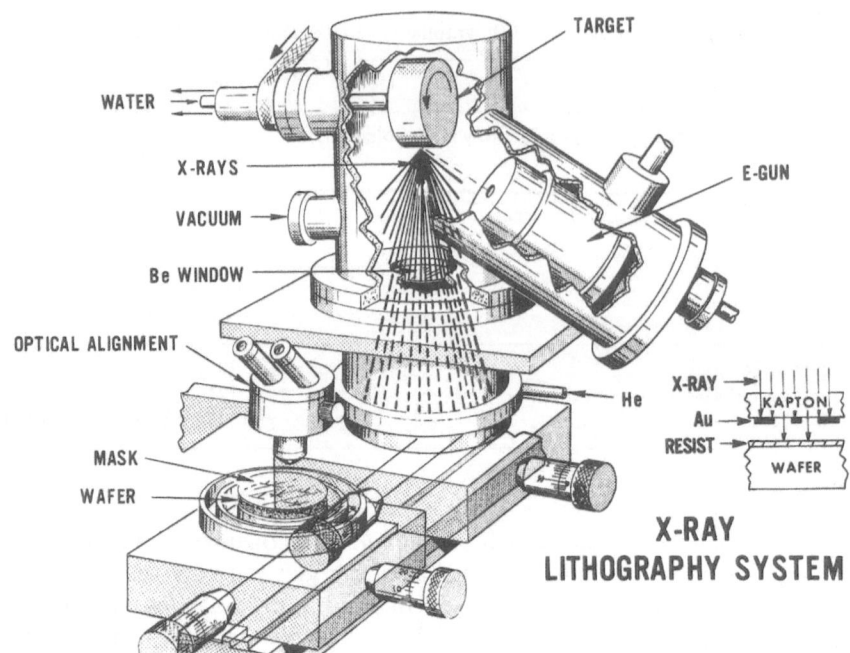

Fig. 3.25. X-ray exposure station with alignment built at Bell Telephone Laboratories (courtesy of *Maydan* [3.38])

Table 3.5 Parameters of the X-ray lithographic system of Fig. 3.25

Anode material	Pd, $\lambda = 4.37\,\text{Å}$
Input power	4.4 kW in 3 mm spot at 25 kV
Vacuum window	50 μm Be
Mask substrate	25 μm Kapton (Polyimid)
Effective transmission of window and substrate	0.7
Source wafer distance	50 cm
X-ray flux at wafer	$4\,\text{mJ cm}^{-2}\,\text{min}^{-1}$
X-ray flux absorbed in PMMA	$0.5\,\text{J cm}^{-3}\,\text{min}^{-1}$
Exposure time for PMMA	1000 min for vertical walls
X-ray flux absorbed in DClPA	$5\,\text{J cm}^{-3}\,\text{min}^{-1}$
Exposure time for DClPA	2 min

0.5 μm which is sufficient for all presently foreseeable silicon device applications. The disadvantage of the system is that it is not easily adaptable to higher resolution. The short wavelengths present in the source spectrum produce contrast problems for Au thicknesses below 0.5 μm [3.42]. Resists of higher resolution will require higher exposures [$E_{\text{inc}} \approx 1/\delta^2$ (3.12)], and the tighter requirement for run-off control will require larger distances between wafer and source which will reduce the power density at the wafer quadratically with the dimensions of the smallest linewidth. The required exposure time, therefore, will increase proportionally to $1/\delta^4$ and will become prohibitively long for substantially smaller device geometries. A one-wafer-exposure system with tight geometrical distortion control would require a synchrotron radiation source for the replication of very small linewidth patterns ($\delta < 0.1$ μm) with short exposure time.

It should be noted that in a system like that shown in Fig. 3.25, the total exposure time per wafer does not change if step and repeat exposures instead of a single full wafer exposure are used. The smaller size of the exposed area allows a smaller source-wafer distance for the same geometrical distortion, thus shortening the exposure time for the individual exposure so that the total exposure time for all exposures is the same as that for a single full wafer exposure [3.26].

3.10 Applications

As of today, X-ray lithography is not yet used commercially to fabricate devices. All uses of X-ray lithography have been restricted to research and development laboratories. Simple devices with uncomplicated processing steps were chosen in the beginning to evaluate this technology. Examples are acoustic surface wave transducers [3.9, 16, 88, 96] and diffraction gratings [3.133, 134].

3.10.1 Magnetic Bubble Devices

Magnetic bubble devices [3.22, 23, 130, 135, 136] seem very well suited for an application of X-ray lithography, because some of the circuits can be fabricated as single level devices [3.137]. Therefore, it is possible to fabricate devices with very small linewidths without the need to first solve all the potential problems of an alignment system. However, the technology of magnetic bubble devices is relatively new and many materials and processing problems have to be solved before one can proceed to the extremes. Although X-ray lithography has demonstrated a resolution better than 100 Å [3.52], the smallest linewidth seriously used for bubble devices is around 1 μm. Figure 3.26 gives an example of the X-ray lithographic fabrication steps of a 1 μm linewidth bubble memory. A master mask with the device pattern is first written by a scanning electron beam system [3.11]. Figure 3.26a shows part of the e-beam mask after exposure and development, (b) after plating and removal of the resist. The master mask in (b) is copied several times in succession with X-rays ($\lambda = 8.3$ Å) to produce several negative masks (c, d) which in turn are replicated simultaneously in a parallel exposure system to produce the device resist pattern (e), and plated device (f). The gold of the mask, the gold in the conductor lines of the device, and the permalloy for the magnetic elements of the bubble memory are all deposited by

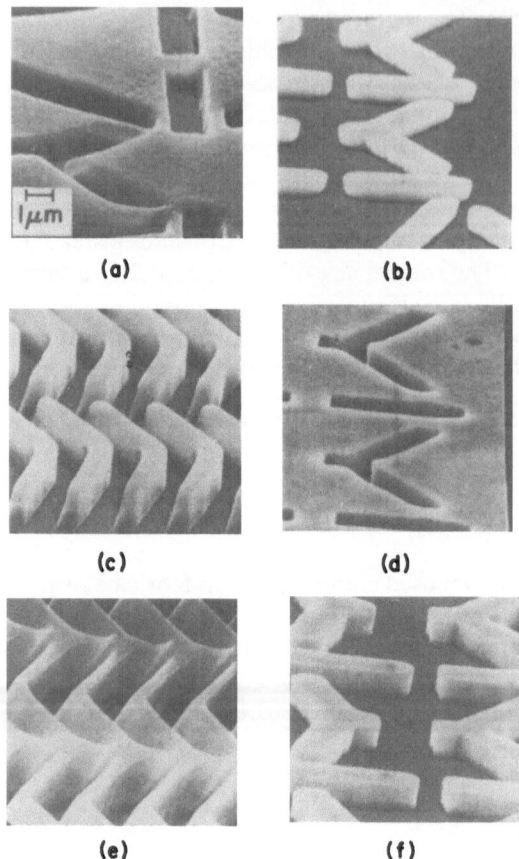

(a)   (b)

(c)   (d)

(e)   (f)

Fig. 3.26a–f. Fabrication steps for a bubble memory using electroplating for all processing steps. X-ray resist patterns (left) and metal patterns after resist removal (right). e-beam fabricated master mask (top), negative X-ray copy (middle) and device pattern obtained from the negative mask (bottom) linewidth 1 μm, SEM micrographs with 60° angle

electroplating; positive resist is used for all exposures. The main advantage of the intermediate mask is the possibility to improve the contrast of the device mask by this step and thus obtain relaxed exposure, development, and processing conditions. Bubble devices have also directly been fabricated from the original e-beam mask and practically all processing steps discussed in Section 3.8 have been utilized [3.23, 38, 130, 136].

3.10.2 Semiconductor Devices

The smallest commercial semiconductor devices have linewidth of 2.5 μm. Devices with 1.5 μm linewidth have been made in the laboratory by e-beam exposures [3.138]. Therefore, semiconductor devices can presently not be used to test the resolution and obtainable yield for submicron devices; tested circuits and processes for these geometries do not exist. Bernacki and Smith [3.20, 89]

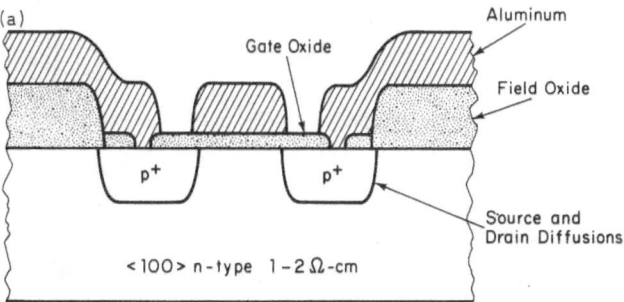

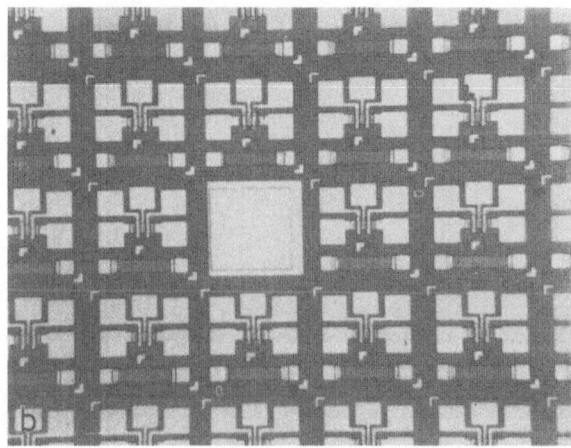

Fig. 3.27. (a) Schematic cross section and (b) optical micrograph of field effect transistors fabricated in an array by X-ray lithography. Oxide custs for source and drain are 2.5 µm by 28 µm, channel region is 5 µm long and 41 µm wide (courtesy of *Smith* [3.89])

used X-ray lithography to produce $p-n$ diodes, bipolar transistors, and MOS transistors with 2.5 µm linewidth (Fig. 3.27). They reported that the initial fabrication run of the devices in a normal laboratory environment had a higher yield (99 percent for 324 MOS transistors) than the fabrication of the same devices in a cleanroom environment with conventional photolithography (90 percent for 780 devices). *Bernacki* and *Smith* could also show that the X-ray induced damage in MOS capacitors (probably trapped charges) could be completely repaired by annealing the devices for 15 minutes at 500 °C in nitrogen.

Silicon devices are also fabricated at Bell Laboratories [3.38] using the exposure system of Fig. 3.25, and again the system is used to fabricate the same devices which are also produced by other lithographic methods to allow a comparison of the different lithographies. The X-ray lithographic system of Fig. 3.25 is designed so that it can directly replace photolithographic exposure systems in an existing manufacturing line and there is a good chance that it will become the first commercial application of X-ray lithography.

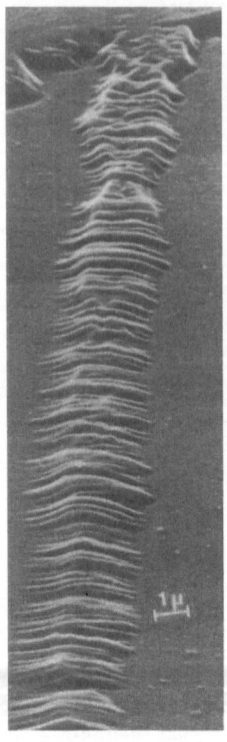

Fig. 3.28. Soft X-ray replica of a part of a chromosome of drosophila in PMMA obtained with carbon K_α-radiation (from [3.52])

3.10.3 Microscopy of Biological Objects

The fast progress in X-ray lithography since 1972 has been applied to contact microscopy and has finally made it possible to obtain a resolution which is outstandingly better than the resolution of an optical microscope and close to the soft X-ray diffraction limit. In X-ray microscopy the specimen is brought in close contact to the resist surface, mounted either on a thin substrate, a grid, or directly on top of the resist layer. After development, a relief structure is obtained where the height of a feature in the resist corresponds to the X-ray absorption of the specimen. A scanning electron microscope produces a magnified picture of the resist relief image. The use of the SEM and of X-ray resists are the key advances since 1972 which allow substantially improved results. The micrographs in Figs. 3.28 and 3.29 demonstrate the state of the art [3.52]. Figure 3.28 shows a part of the resist relief structure obtained from a chromosome from the salivary glands of drosophila. Figure 3.29 is a replica of pigment epithelium granules from the retina of the frog, Rana catesbeiana. The resolution in Fig. 3.29 is better than that of a conventional scanning electron microscope, and a special high resolution SEM which uses only the electrons

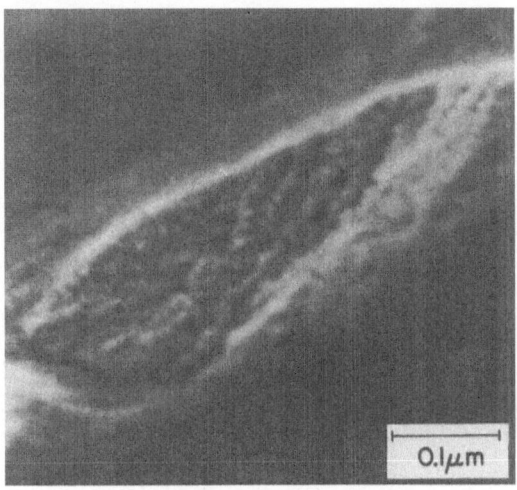

Fig. 3.29. Soft X-ray replica of a pigment granule from frog retina obtained with synchrotron radiation (from [3.52])

with little energy loss is used to bring out the smallest details in the topography of the resist surface. The details in Figs. 3.28 and 3.29 have not been seen by any other microscopic technique and many interesting biological applications can be expected. In addition to the capability of revealing an internal structure of a specimen with high resolution, X-ray microscopy has also the potential of obtaining this information with less radiation damage to the specimen than with an electron microscope [3.139]. The lower damage level is due to the higher contrast which specimens have for soft X-rays compared to that for high energy electrons. Live specimens can be more easily observed with soft X-rays than with electron beams. Synchrotron radiation sources are the ideal sources for soft X-ray microscopy because they allow short exposure even for the highest resolution.

The work in X-ray microscopy has demonstrated that X-ray lithography has a resolution capability around $\delta = 50\,\text{Å}$ and that a resist like PMMA has sufficient contrast to replicate such small structures. Structures of this and of smaller dimensions are abundant in biological specimens, while the electronic circuits still have a long way in their development before such small features can be considered.

3.11 Performance and Limitations

In this section we shall summarize the main advantages and disadvantages of X-ray lithography compared to other techniques. Many advantages of X-rays are due to the fact that X-rays interact only weakly with matter. The refractive index

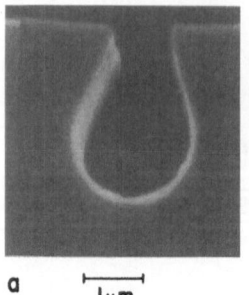
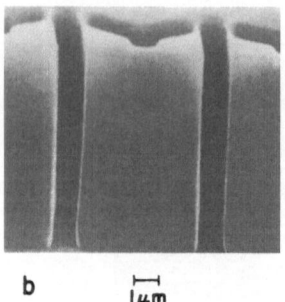

a |—1 μm—| b |—1 μm—|

Fig. 3.30a and b. Resist patterns obtained in thick resist with (a) electrons of 15 kV energy (from *Hatzakis* [3.108]) and (b) with X-rays of ∼2.5 keV ($\lambda = 4.5$ Å) energy. Both patterns have a linewidth of 1 μm at the surface of the resist

of all materials is practically one, therefore, X-rays are practically not refracted or scattered by matter. The wavelength of soft X-rays is much shorter than the dimensions of all devices presently considered so that diffraction effects can be neglected in most cases.

For all practical purposes X-rays travel in straight lines through matter. A consequence is that a pattern geometry is maintained throughout thick layers of resist material; patterns can be generated with *high aspect ratios* (height to width ratios). Figure 3.30 shows a 1 μm linewidth patterns in thick resist obtained by (a) *e*-beam exposure and (b) by X-ray exposure; the wafer has been broken after development to take the SEM micrograph. Electrons are scattered in resist and have a scattering range of several μm for the energies used in *e*-beam systems. This scattering produces resist profiles which depend critically on exposure and development conditions [3.107, 108]. Figure 3.30a was obtained with a resist film so thick that no scattering from the substrate occurs. In practice the backscattering from the substrate makes the effect even worse than Fig. 3.30a. Electrons can expose resist away from their original impact point. These proximity effects [3.109] require a delicate local adjustment of the exposure and tight control of the development for acceptable results. The high resolution capability of an *e*-beam system can be utilized only for pattern writing, if extremely thin substrates (∼200 Å thick carbon) and resist films are used [3.140]; in a normal system, electron scattering is an annoying factor even for relatively coarse structures in the 1 μm range. The high aspect ratio obtainable with X-rays (Fig. 3.30b, resist thickness 14 μm) cannot be obtained by any other lithographic technique with comparable resolution.

Loose tolerances can be expected for X-ray lithography for linewidths above 0.1 μm with masks of sufficient contrast, provided that a resist of sufficiently high resolution is used. Developed resist walls become steeper and steeper for higher and higher exposures. A certain minimum exposure is required for nearly vertical walls; above that exposure the shape and slope of the resist wall do not depend on the exposure (Fig. 3.13). The loose tolerances are useful to allow X-ray lithography to be *insensitive to dust*. Most laboratory dust is organic and is very transparent to X-rays. We can simply ignore all dust particles on the mask which for example absorbs less than 50 percent of the incident X-rays (corresponding

Fig. 3.31. Elimination of a 10 μm thick string of glue on an X-ray mask by high exposure and prolonged development. Linewidth in the patterns is 1 μm, (a) mask, (b) replica in resist with arrows marking one edge of the glue string replica

a   b

to particle sizes up to 5 μm for organic dust and $\lambda = 8.3$ Å) by increasing the exposure by a factor 2 to give still sufficient exposure to the parts of the resist which are shadowed by the dust particle. Figure 3.31 is an example in which this method has been used to eliminate the replication of a 10 μm thick string of glue which was deliberately put on the mask. The insensitivity to organic dust is probably responsible for the high yield obtained in the fabrication of semiconductors [3.89].

Mask life appears to be another advantage of X-ray lithography. Mask and wafer never have to brought into contact due to the large depth of focus. Therefore, mask life is limited by either accidents or radiation damage. The latter allows at least 10^4 replications for a mylar mask in PMMA resist [3.50] and probably even more for the inorganic mask substrates. Deep UV printing has demonstrated impressive aspect ratios in resist exposure [3.141]; however, it cannot compete with X-ray lithography in mask life, insensitivity to dust, and processing tolerances.

The *low efficiency* ($\sim 10^{-4}$) of conventional X-ray sources is a disadvantage for X-ray lithography. Problems in throughput have to be expected for high resolution work (linewidth ≈ 0.1 μm) in systems which expose only one wafer at a time. Synchrotron radiation sources can eliminate this problem; however, cost and overcapacity which might require that one source be shared by several users are big disadvantages.

Precision alignment (better than 0.1 μm) has not yet been demonstrated for device fabrication. Wafer and mask stability through all processing levels may become the fundamental limit for the commercial success of further device miniaturization. All lithographic techniques which use a fixed mask will suffer from the same problem and only computer controlled scanning (*e*-beam) systems which can measure and correct for local wafer distortions are free from it. The *first mask* for X-ray lithography has to be produced by an *e*-beam system. Therefore, X-ray lithography is also affected by the disadvantages of an *e*-beam system (electron scattering and proximity effects). The fabrication of the first mask will be an elaborate, time consuming, low yield process when linewidths close to 0.1 μm are required. Mask production with soft X-rays is much easier, but would require X-ray focusing elements.

3.12 Outlook: Focusing Elements for Soft X-Rays

The challenge to build a magnifying soft X-ray microscope is still open for future work. While it might appear that the high resolution obtained with contact microscopy in combination with an SEM has made such an instrument unnecessary—one cannot hope to further improve the resolution drastically—there are still interesting applications for such an instrument. Figure 3.32 shows a scanning X-ray microscope which could be built if a focusing element for X-rays were available. The small focal spot scans the specimen and the transmission values are measured by a detector behind the object and stored in a memory or used for display. This system would have the following advantages over 1:1 contact microscopy:

1) The detector (photon counter) could have a very high quantum efficiency. The performance would be limited only by the shot noise as discussed in Section 3.6 and would allow observation with the minimum radiation damage to the specimen compatible with the resolution and signal/noise requirements.

2) Quantitative analysis of the object structure would be much easier than with the X-ray resist technique. As an example, let us assume that we want to obtain a microscopic image of the calcium distribution in the specimen. A subtraction of two images in the memory, one obtained with X-ray wavelengths slightly above, and the other with wavelengths slightly below a calcium absorption edge, could immediately provide the desired result. It would be very time consuming and laborious to obtain the same result with the resist method.

3) The microscope in Fig. 3.32 could also be used as a pattern generator which could write the first X-ray mask. The object in Fig. 3.32 would be the resist-coated mask substrate, the x, y position is determined by a computer program which also controls a shutter in the beam line to produce the desired pattern. The construction of this machine would be very similar to that of a scanning *e*-beam system [3.11]. Exposure by X-rays instead of *e*-beams would eliminate all the problems connected with the scattering of electrons and promises a higher reliability for the mask fabrication process.

The microscope in Fig. 3.32 requires synchrotron radiation for sufficient brightness in the spot and a focusing element capable of high resolution.

Fresnel zone plates have been used as X-ray focusing elements for many years [3.142–145]. Recently a resolution of about 0.5 µm has been obtained with X-ray wavelengths around 50 Å [3.146]. Theoretically, the resolution of a zone plate is given by the width of the finest zones. Zone plates have been produced either holographically [3.144]—in this case, the shortest available wavelength for producing the hologram limits the resolution—or by electron beams [3.142, 143]. Lines with 80 Å width have recently been fabricated with a scanning *e*-beam system [3.140]. The limit in the performance of a scanning *e*-beam fabricated zone plate will not, in practice, however, be determined by this resolution but by the stability of the *e*-beam during the time required to write the

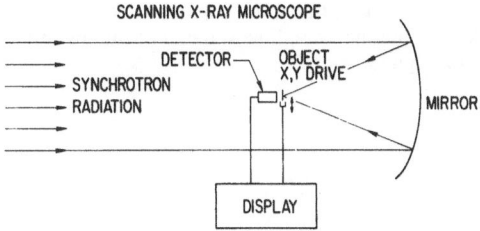

Fig. 3.32. Scanning X-ray microscope

entire zone plate and by subsequent processing steps. X-ray lithography can be used to improve the contrast of the first zone plate, and simultaneously (by the use of a divergent beam) might produce a blazed zone plate with high diffraction efficiency for the first order.

Grazing incidence focusing mirrors have been used in X-ray microscopes and telescopes [3.5, 147, 148]. The aberrations of these instruments and the difficulties in producing the designed surfaces have, however, limited the resolution. *Normal incidence mirrors* could have substantially better performance; however, the reflectivity obtainable from conventional materials is too low for any practical application ($R < 10^{-5}$ at $\lambda = 50$ Å). Multilayer coatings can enhance the normal incidence reflectivity of a surface in the soft X-ray region to values about 30 percent [3.149, 150], a value high enough to make them useful in a microscope as shown in Fig. 3.32. Optimized, aperiodic multilayer reflectors have been made by evaporation for wavelengths around $\lambda = 2000$ Å [3.151] and $\lambda = 200$ Å [3.152], and attempts are now under way to fabricate practical useful normal incidence reflectors in the 50–200 Å wavelength region. Periodic multilayer coatings of lead-fatty acids and similar compounds produced by the Langmuir-Blodgett method have up to now been used only as spectral analyzers in the soft X-ray region [3.30]. They might also be useful as reflective coating for a focusing mirror.

There is a good chance that within the next years one of the discussed methods will finally make high quality focusing elements for soft X-rays possible. These new elements would add much more flexibility to the present applications of the soft X-ray region and certainly make new ones possible.

Acknowledgments. Our work on X-ray lithography at the IBM Research Laboratory was based on the technical expertise of many people.

We thank *T. H. P. Chang* and *H. Luhn* for their support in the *e*-beam fabrication of masks, *K. Ahn, M. Blakeslee, E. Castellani, M. Heritage, S. Kane, S. Krongelb, Y. Powers, L. T. Romankiw, T. Tuxford* and *C. B. Zarowin* for contributions to the processing of masks and devices, *I. Haller* and *M. Hatzakis* for their work on resists, *E. Munro* and *G. Wardly* for the design and construction of a high power X-ray source, *A. N. Broers* and *D. Sayre* for their cooperation in microscopy, *D. Eastman, W. Grobman* and *W. Gudat* for help in the use of synchrotron radiation, and *G. Grant* for the preparation of the manuscript. The experiments with synchrotron radiation were performed at the DESY synchrotron in Hamburg and we thank the staff and management of DESY for their hospitality and support. The reliable technical help of *J. Topalian* in practically all experiments was very important to all phases of our work.

We also thank *H. I. Smith* (MIT) and *D. Mayden* (Bell Laboratories) and *A. R. Neureuther* (Berkeley) for providing us freely with information about the work done at their institutions and for the permission to reprint several figures from their publications.

References

3.1   P. Goby: C. R. Acad. Sci. Paris **156**, 686 (1913)

3.2   W. C. Nixon: Proc. Roy. Soc. A **232**, 475 (1955)

3.3   W. A. Ladd, M. Hess, M. W. Ladd: Science **123**, 370 (1956)

3.4   S. K. Assunmaa: *X-Ray Optics and X-Ray Microanalysis*, ed. by H. H. Patte, V. E. Cosslett, A. Engstrom (Academic Press, New York 1963) p. 33

3.5   P. Kirkpatrik, H. H. Patte, Jr.: *X-Ray Microscopy*, in Handbuch der Physik, ed. by S. Flügge, Vol. XXX (Springer, Berlin, Heidelberg, New York 1957)

3.6   V. E. Cosslett, W. C. Nixon: *X-Ray Microscopy* (University Press, Cambridge 1960)

3.7   T. A. Hall, H. O. E. Röckert, R. L. de C. H. Saunders: *X-Ray Microscopy in Clinical and Experimental Medicine* (C. T. Thomas, Springfield, Ill. 1972)

3.8   D. L. Spears, H. I. Smith: Electron. Letters **8**, 102 (1972)

3.9   D. L. Spears, H. I. Smith: Solid State Technol. **15**, 21 (1972)

3.10   D. R. Herriot, R. J. Collier, D. S. Alles, J. W. Stafford: IEEE Trans. ED-**22**, 385 (1975)

3.11   T. H. P. Chang, A. D. Wilson, A. J. Speth, A. Kern: *Electron and Ion Beam Science and Technology*, 6th Intern. Conf., ed. by R. Bakish (Electrochemical Society, Princeton, N.J. 1974), p. 580

3.12   G. L. Clark (ed.): *The Encyclopedia of X-Rays and Gamma Rays* (Reinhold, New York 1973), pp. 11–15

3.13   B. L. Bracewell, W. J. Veigele: *Developments in Applied Spectroscopy*, ed. by E. L. Grove, A. J. Perkins, Vol. 9 (Plenum Press, New York 1971), p. 375

3.14   B. L. Henke, E. S. Ebisu: *Advances in X-Ray Analysis*, Vol. 17 (Plenum Press, New York 1973), p. 150–213

3.15   H. J. Hagemann, W. Gudat, C. Kunz: J. Opt. Soc. Am. **65**, 742 (1975)

3.16   D. L. Spears, H. I. Smith, E. Stern: *Electron and Ion Beam Science and Technology*, 5th Intern. Conf., ed. by R. Bakish (Electrochemical Soc., Princeton, N.J. 1972), p. 80

3.17   R. A. Cohen, R. W. Mountain, D. L. Spears, H. I. Smith, M. A. Lemma, S. E. Bernacki: "Fabrication procedure for silicon membrane X-ray lithography masks"; M.I.T. Lincoln Lab., Lexington, Mass. 02173, Tech. Note 1973-38, Sept. 20, 1973, AD-769857/4

3.18   H. I. Smith, S. E. Bernacki: J. Vac. Sci. Technol. **12**, 1321 (1975)

3.19   J. H. McCoy, P. A. Sullivan: *Electron and Ion Beam Science and Technology*, 6th Intern. Conf., ed. by R. Bakish (Electrochem. Soc., Princeton, N.J. 1974), p. 3

3.20   S. E. Bernacki, H. I. Smith: *Electron and Ion Beam Science and Technology*, 6th Intern. Conf., ed. by R. Bakish (Electrochem. Soc., Princeton, N.J. 1974), p. 34

3.21   C. J. Schmidt, P. V. Lenzo, E. G. Spencer: J. Appl. Phys. **46**, 4080 (1975)

3.22   E. Spiller, R. Feder, J. Topalian, E. Castellani, L. Romankiw, M. Heritage: Solid State Technol. **19** No. 4, 62 (1976)

3.23   D. Mayden, G. A. Coquin, J. R. Maldonado, S. Somekh, D. Y. Lou, G. N. Taylor: IEEE Trans. ED-**22**, 429 (1975)

3.24   J. S. Greeneich: IEEE Trans. ED-**22**, 434 (1975)

3.25   P. A. Sullivan, J. H. McCoy: J. Vac. Sci. Technol. **12**, 1325 (1975)

3.26   H. I. Smith, D. C. Flanders: Jap. J. Appl. Phys. **16**, Suppl. **16-1**, 61 (1977)

3.27   E. Bassous, R. Feder, E. Spiller, J. Topalian: Solid State Technol. **19**, No. 9, 55 (1976)

3.28   D. C. Flanders, H. I. Smith: 14th Symposium on Electron Ion and Photon Beam Technology (Palo Alto, Calif., 1977) to be published J. Vac. Science Technol.

3.29   T. Funayama, Y. Takayama, T. Inagaki, M. Nakamura: J. Vac. Sci. Technol. **12**, 1324 (1975)

3.30 B.L.Henke, M.A.Tester: *Advances in X-Ray Analysis*, ed. by W.L.Pickles, C.S.Barrett, J.B.Newkirk, C.O.Rund, Vol. 18 (Plenum Press, New York 1974), p. 76
3.31 D.M.Barrus, R.L.Blake: Rev. Sci. Instr. **48**, 116 (1977)
3.32 M.J.Schwartz: *Electro-Optical Systems Design* **2** No. 8, 88 (1970)
3.33 M.A.Spivack: Rev. Sci. Instr. **41**, 1614 (1970)
3.34 D.W.Havas, R.S.Horwath: Extended Abstracts, No. 278, Fall Meet (Electrochem. Soc., Miami, Fla. 1972)
3.35 G.Wardly, E.Munro, R.W.Scott: Intern. Conf. on Microlithographie (Paris 1977), p. 217
3.36 R.M.Dolby: Brit. J. Appl. Phys. **11**, 64 (1960)
3.37 G.N.Taylor, G.A.Coquin, S.Somekh: 4th Internatl. Techn. Conf. on Photopolymers (Ellenville, N.Y. 1976), Polymer Engineering and Science **17**, 420 (1977)
3.38 D.Maydan, G.A.Coquin, J.R.Maldonado, J.M.Moran, S.Somekh, G.N.Taylor: International Conference on Photolithography (Paris 1977), p. 195
3.39 R.Feder, E.Spiller, J.Topalian, M.Hatzakis: *Electron and Ion Beam Science and Technology*, 7th Intern. Conf., ed. by R.Bakish (Electrochem. Soc., Princeton, N.J. 1976), p. 198
3.40 E.Spiller, R.Feder, J.Topalian, D.Eastman, W.Gudat, D.Sayre: Science **191**, 1172 (1976)
3.41 P.A.Sullivan, J.H.McCoy: IEEE Trans. ED-**23**, 412 (1976)
3.42 J.R.Maldonado, G.A.Coquin, D.Maydan, S.Somekh: J. Vac. Sci. Technol. **12**, 1329 (1975)
3.43 R.Feder, E.Spiller, J.Topalian: J. Vac. Sci. Technol. **12**, 1332 (1975)
3.44 E.Hundt, P.Tischer: 14th Symposium on Electron, Ion and Photon Beam Technology (Palo Alto, Calif. 1977) to be published J. Vac. Science Technol.
3.45 M.L.Perlman, E.M.Rowe, R.E.Watson: Phys. Today **27**, 30 (1974)
3.46 E.Koch, C.Kunz: Topics in Appl. Phys. (to be published)
3.47 C.Kunz: Physik Bl. **32**, 9, 55 (1976)
3.48 P.Dagneaux, C.Depautex, P.Dhez, J.Durup, Y.Farge, R.Fourme, P.M.Guyon, P.Jaegle, S.Leach, R.Lopez-Delgado, G.Morel, R.Pinchaux, P.Thiry, C.Vermeil, F.Wuilleumier: Ann. Phys. (Paris) **9**, 9 (1975)
3.49 C.Kunz: *Vacuum Ultraviolet Radiation Physics*, ed. by E.E.Koch, R.Haensel, C.Kunz (Vieweg, Braunschweig 1974), p. 753
3.50 E.Spiller, D.E.Eastman, R.Feder, W.D.Grobman, W.Gudat, J.Topalian: J. Appl. Phys. **47**, 5450 (1976)
3.51 B.Fay, J.Trotel: 14th Symposium on Electron, Ion and Photon Beam Technology (Palo Alto, Calif. 1977) to be published J. Vac. Science Technol.
3.52 R.Feder, E.Spiller, J.Topalian, A.N.Broers, W.Gudat, B.J.Panessa, Z.A.Zadunaisky, J.Sedat: Science **197**, 259 (1977)
3.53 B.Fay, J.Trotel, Y.Petroff, R.Pinchaux, P.Thiry: Appl. Phys. Letters **29**, 370 (1976)
3.54 J.Trotel, B.Fay: Intern. Conf. on Microlithography (Paris 1977), p. 201
3.55 T.W.O'Keefe, R.M.Handy: Solid State Electron. **11**, 261 (1968)
3.56 B.H.Hill: J. Electrochem. Soc. **116**, 66 (1969)
3.57 W.R.Sinclair, D.L.Rousseau, J.J.Stancavish: J. Electrochem. Soc. **121**, 925 (1974)
3.58 G.W.Kammlott, W.R.Sinclair: J. Electrochem. Soc. **121**, 929 (1974)
3.59 A.N.Wright: U.S. Patent 3664899
3.60 R.Feder, F.P.Laming, E.Spiller, J.Topalian: IBM Tech. Discl. Bull. **19**, 316 (1976)
3.61 D.E.Eastman, R.Feder, W.Gudat, E.Spiller, J.Topalian: IBM Techn. Discl. Bull. (to be published)
3.62 W.J.McGill: J. Appl. Polymer Science **19**, 2781 (1972)
3.63 M.J.Bowden, L.F.Thompson: J. Appl. Polymer Sci. **17**, 3211 (1973)
3.64 H.Hiraoka: IBM J. Res. Devel. **21**, 121 (1977)
3.65 L.F.Thompson, R.E.Kerwin: *Annual Review of Materials Science*, ed. by R.A.Huggins, R.H.Babe, R.W.Roberts, Vol. 6 (Annual Reviews, Inc., Palo Alto, Calif. 1976), p. 267
3.66 L.H.Princen (ed.): *Scanning Electron Microscopy of Polymers and Coatings II.* Applied Polymer Symposia 23 (J. Wiley, New York 1974)
3.67 A.Charlesby: *Atomic Radiation and Polymers* (Pergamon Press, London 1960)
3.68 W.S.DeForest: *Photoresist, Materials and Processes* (McGraw-Hill, New York 1975)

3.69   A.Papoulis: *Probability, Random Variables and Stochastic Processes* (McGraw-Hill, New York 1965)
3.70   I.Haller, M.Hatzakis, R.Srinivasan: IBM J. Res. Devel. **12**, 251 (1968)
3.71   J.S.Greeneich: J. Electrochem. Soc. **121**, 1669 (1974)
3.72   M.Hatzakis, C.H.Ting, N.Viswanathan: *Electron and Ion Beam Science and Technology*, 6th Intern. Conf., ed. by R.Bakish (Electrochem. Soc., Princeton, N.J. 1974), p. 542
3.73   M.Hatzakis: J. Electrochem. Soc. **116**, 1033 (1969)
3.74   R.Feder, D.Sayre, E.Spiller, J.Topalian, J.Kirz: J. Appl. Phys. **47**, 1192 (1976)
3.75   R.Feder, I.Haller, M.Hatzakis, L.T.Romankiw, E.Spiller: U.S. Patent 3984582
3.76   P.I.Hagouel, A.R.Neureuther: *Electron and Ion Beam Science and Technology*, 7th Intern. Conf., ed. by R.Bakish (Electrochem. Soc., Princeton, N.J. 1976), p. 190
3.77   P.I.Hagouel, A.R.Neureuther: ACS Organic Coating and Plastics Preprints **35**, No. 2, 298 (1975)
3.78   R.E.Jewett, P.I.Hagouel, A.R.Neureuther, T. Van Duzer: 4th Intern. Techn. Conference on Photopolymers (Ellenville, N.Y. 1976), Polymer Engineering and Science **17**, 381 (1977)
3.79   A.R.Neureuther: 14th Symposium on Electron, Ion and Photon Beam Technology (Palo Alto, Calif. 1977)
3.80   F.H.Dill, A.R.Neureuther, J.A.Tuttle, E.J.Walker: IEEE Trans. ED-**22**, 456 (1975)
3.81   R.Feder, E.Spiller, J.Topalian: 4th Intern. Techn. Conf. on Photopolymers (Ellenville, N.Y. 1976), Polymer Engineering and Science **17**, 385 (1977)
3.82   R.G.Brault: *Electron and Ion Beam Science and Technology*, 6th Intern. Conf., ed. by R.Bakish (Electrochem. Soc., Princeton, N.J. 1974), p. 63
3.83   L.F.Thompson, E.D.Feit, M.J.Bowden, P.V.Lenzo, E.G.Spencer: J. Electrochem. Soc. **121**, 1500 (1974)
3.84   G.N.Taylor, G.A.Coquin: private communication
3.85   R.Feder, I.Haller, M.Hatzakis, L.T.Romankiw, E.Spiller: IBM Techn. Discl. Bull. **18**, 2346 (1975)
3.86   R.Feder, I.Haller, M.Hatzakis, L.T.Romankiw, E.Spiller: IBM Techn. Discl. Bull. **18**, 2343 (1975)
3.87   P.V.Lenzo, E.G.Spencer: Appl. Phys. Letters **24**, 289 (1974)
3.88   H.I.Smith, D.L.Spears, S.E.Bernacki: J. Vac. Sci. Technol. **10**, 913 (1973)
3.89   S.E.Bernacki, H.I.Smith: IEEE Trans. ED-**22**, 421 (1975)
3.90   T.Hirai, Y.Hatano, S.Nonogaki: J. Electrochem. Soc. **118**, 669 (1971)
3.91   K.G.Clark: Solid State Technology **14**, No. 2, 48 (1971)
3.92   M.C.King, D.H.Berry: Appl. Opt. **11**, 2455 (1972)
3.93   K.G.Clark, K.Okutsu: Solid State Technology **19**, No. 4, 79 (1976)
3.94   D.A.Markle: Solid State Technology **17**, No. 6, 50 (1974)
3.95   J.Schwider, C.H.Hiller: Optica Acta **23**, 49 (1976)
3.96   H.I.Smith: Proc. IEEE **62**, 1361 (1974)
3.97   J.H.McCoy, P.S.Sullivan: Solid State Techn. **19**, No. 9, 59 (1976)
3.98   J.H.McCoy, P.A.Sullivan: *Electron and Ion Beam Science and Technology*, 7th Intern. Conf., ed. by R. Bakish (Electrochem. Society, Princeton, N.J. 1976), p. 536
3.99   S.Yamazaki, S.Nakayama, T.Hayasaka, S.Ishihara: 14th Symposium on Electron, Ion and Photon Beam Technology (Palo Alto, Calif. 1977) to be published J. Vac. Science Technol.
3.100 S.Austin, D.C.Flanders, H.I.Smith: 14th Symposium on Electron, Ion and Photon Beam Technology (Palo Alto, Calif. 1977) to be published J. Vac. Science Technol.
         D.C.Flanders, H.I.Smith, S.Austin: Appl. Phys. Letters **31**, 426 (1977)
3.101 A.D.White: Appl. Opt. **16**, 549 (1977)
3.102 M.Kallmeyer, K.Kosanke, F.Schedewie, B.Solf, D.Wagner: IBM J. Res. Develop. **17**, 490 (1973)
3.103 E.Hundt, P.Tischer: International Conference on Photolithography (Paris 1977), p. 211
3.104 D.E.Eastman, R.Feder, W.Grobman, E.Spiller, J.Topalian: IBM Techn. Discl. Bull. **18**, 3111 (1976)

3.105 R.Feder, P.Garbarino, C.Johnson, E.Spiller, J.Topalian: IBM Techn. Discl. Bull. **19**, 4441 (1977)
3.106 R.C.Henderson, R.F.W.Pease: Polymer Eng. Sci. **14**, 538 (1974)
3.107 J.S.Greeneich: J. Appl. Phys. **45**, 5264 (1974)
3.108 M.Hatzakis: Appl. Phys. Letter **18**, 7 (1971)
3.109 T.H.P.Chang: J. Vac. Sci. Technol. **12**, 1271 (1975)
3.110 R.Feder, M.Hatzakis, M.Heritage, E.Spiller: IBM Techn. Discl. Bull. **17**, 2460 (1975)
3.111 R.Feder, M.B.Heritage, E.Spiller, J.Topalian: IBM Techn. Discl. Bull. **18**, 3110 (1975)
3.112 A.D.Wilson, T.H.P.Chang, A.Kern: J. Vac. Sci. Technol. **12**, 1240 (1975)
3.113 D.S.Alles, F.R.Ashley, A.M.Johnson, R.L.Townsend: J. Vac. Sci. Technol. **12**, 1252 (1975)
3.114 A.D.Wilson, A.Kern, A.J.Speth, A.M.Patlach, P.R.Jaskar, T.L.Keller: *Electron and Ion Beam Science and Technology*, 7th Intern. Conf., ed. by R. Bakish (Electrochem. Society, Princeton, N.J. 1976), p. 361
3.115 H.G.Hughes, M.J.Rand (ed.): *Etching for Pattern Definition* (Electrochem. Society, Princeton, N.J. 1976)
3.116 W.Kern: Ref. [3.115] p. 1
3.117 L.T.Romankiw: Ref. [3.115] p. 161
3.118 D.MacArthur: Ref. [3.115] p. 76
3.119 A.R.Reinberg: Ref. [3.115] p. 91
3.120 H.I.Smith: Ref. [3.115] p. 133
3.121 D.C.Flanders, H.I.Smith: 14th Symposium on Electron, Ion and Photon Beam Technology (Palo Alto, Calif. 1977)
3.122 R.G.Poulsen: J. Vac. Sci. Technol. **14**, 266 (1977)
3.123 D.T.Hawkins: J. Vac. Sci. Technol. **12**, 1389 (1975)
3.124 S.Somekh, H.C.Casey: J. Appl. Opt. **16**, 126 (1977)
3.125 H.W.Lehmann, L.Krausbauer, R.Widmer: J. Vac. Sci. Technol. **14**, 281 (1977)
3.126 W.R.Hudson: J. Vac. Sci. Technol. **14**, 286 (1977)
3.127 M.Hatzakis: Applied Polymer Symposium No. 23 (J. Wiley, New York 1974), p. 73
3.128 L.T.Romankiw, S.Krongelb, E.E.Castellani, B.J.Stoeber, J.D.Olsen: IEEE Trans. MAG-**10**, 828 (1974)
3.129 K.A.Pickar: *Appl. Solid State Science* 5, ed. by R.Wolfe (Academic Press, New York 1975)
3.130 T.Funayama, K.Yanagida, N.Kanoyama, K.Kemeno, T.Inagaki: 14th Symposium on Electron, Ion and Photon Beam Technology (Palo Alto, Calif. 1977)
3.131 W.D.Buckley: *Electron and Ion Beam Science and Technology*, 7th Intern. Conf., ed. by R.Bakish (Electrochem. Society, Princeton, N.J. 1976), p. 453
3.132 G.P.Hughes: 14th Symposium on Electron, Ion and Photon Beam Technology (Palo Alto, Calif. 1977) and Solid State Technology **20**, No. 5, 39 (1977)
3.133 A.R.Neureuther, P.I.Hagouel: *Electron and Ion Beam Science and Technology*, 6th Intern. Conf., ed. by R. Bakish (Electrochem. Society, Princeton, N.J. 1974), p. 23
3.134 P.I.Hagouel: X-Ray Fabrication of Blazed Diffraction Gratings, Thesis (University of California, Berkeley 1976)
3.135 R.K.Watts, H.M.Darley, J.B.Kruger, T.G.Blocker, D.C.Guterman, J.T.Carlo, D.C.Bullock, M.S.Shaikh: Appl. Phys. Letters **28**, 355 (1976)
3.136 R.K.Watts, D.C.Guterman, H.M.Darley: SPIE **80**, 100 (1976)
3.137 A.H.Bobeck, I.Danylchuck, F.C.Rossol, W.Strauss: IEEE Trans. MAG-**9**, 474 (1973)
3.138 H.N.Yu, R.H.Dennard, T.H.P.Chang, C.M.Osburn, V.Dilonardo, H.E.Luhn: J. Vac. Sci. Technol. **12**, 1297 (1975)
3.139 D.Sayre, J.Kirz, R.Feder, D.M.Kim, E.Spiller: Science (to be published)
3.140 A.N.Broers, W.W.Molzen, J.J.Cuomo, N.D.Wittels: Appl. Phys. Lett. **29**, 596 (1976)
3.141 B.J.Lin: J. Vac. Sci. Technol. **12**, 1321 (1975)
3.142 G.Möllenstedt, H.J.Einighammer, K.H.V.Grote, U.Mayer: *X-Ray Optics and X-Ray Microanalysis* (Hermann, Paris 1966), p. 15

3.143 H. Bräuninger, H. J. Einighammer, H. H. Fink: *X-Ray Optics and Microanalysis* (University of Tokyo Press 1972), p. 17

3.144 G. Schmahl, D. Rudolph: Optik **29**, 577 (1969)

3.145 J. Kirz: J. Opt. Soc. Am. **64**, 301 (1974)

3.146 B. Niemann, D. Rudolph, G. Schmahl: Appl. Opt. **15**, 1882 (1976)

3.147 J. H. Underwood, J. E. Milligan, A. C. deLoach, R. B. Hoover: Appl. Opt. **16**, 859 (1977)

3.148 H. Wolter: Ann. Physik **6**, 94 (1952); **6**, 286 (1952)

3.149 E. Spiller: Appl. Opt. **15**, 2333 (1976)

3.150 A. V. Vinogradov, B. Ya. Zeldovich: Appl. Opt. **16**, 89 (1977)

3.151 E. Spiller: Optik **39**, 118 (1973)

3.152 R.-P. Haelbich, C. Kunz: Opt. Commun. **17**, 187 (1976)

3.153 P. J. Mallozi, H. M. Epstein, R. G. Jung, D. C. Appelbaum, B. P. Fairand, W. J. Gallagher, R. L. Uecker, M. C. Muckerheide: J. Appl. Phys. **45**, 1891 (1974)

3.154 P. J. Mallozi, B. P. Fairand, M. J. Gollis: *Laser Produced X-Rays, Neutrons and Ultrasound, in Research Techniques and Nondestructive Testing*, ed. by R. S. Sharpe (Academic Press, London 1977)

3.155 P. J. Mallozi (private communications)

4. X-Ray and Neutron Interferometry

U. Bonse and W. Graeff

With 25 Figures

The invention of X-ray interferometry by *Bonse* and *Hart* [4.1] has rekindled interest in extending the powerful methods of light interferometry down to the angström range of wavelengths, i.e., to electron waves, X-rays, and thermal neutrons. These radiations are scattered by atoms and nuclei through a variety of interaction processes and are thus unique probes for studying matter on the atomic and—to some extent— subatomic scale. It is hoped, therefore, that the possibly successful adaptation of the methods of interferometry and, maybe, of holography, which in optics are renowned for their directness, abundance of information and accuracy, may substantially promote the study of matter.

The review presented here is partly motivated by the work in this field which has been performed so far with X-rays and recently also with neutrons [4.2].

X-ray interferometry has proved to be a very sensitive tool for the study of defects in nearly perfect crystals. The technique was applied to the precise measurement of atomic scattering factors and the absolute lattice parameter measurement. Although this technique is suited also to phase topography, little has been done in that direction so far. Neutron interferometry has been applied to the precise measurement of coherent scattering lengths b_c, to neutron spin rotation in a magnetic field, to the action of gravity on interfering neutrons, and to the scattering of neutrons by magnetic materials.

Further motivation arises from what may be possible with X-ray and neutron interferometry in the future, once the discovery period is followed by a period of somewhat more systematic study which we hope a review will help to initiate.

With X-rays and in particular with neutrons, fairly thick samples can be investigated. With X-rays, one can study the interaction not only with the electronic shell but possibly also with the nuclear charge because of the accuracy obtainable, especially if the interferometric method is combined with the Pendellösung method [4.3]. Neutrons interact with atoms through the nuclear interaction, the magnetic interaction with magnetic atoms, the relativistic interaction between the electric charges in the atom and the magnetic moment of the neutron (*Foldy* [4.4]), and the *Schwinger* interaction [4.5], just to name those which have been experimentally verified. The last interactions are, however, smaller by a factor of about 10^4. For the measurement of the contribution of the various interactions in the scattering process, an accurate method is very desirable. It appears that neutron interferometry can be developed to sufficient precision to make such measurements possible. *Maier-Leibnitz* and *Springer*

[4.6] pointed out a number of problems in neutron physics the solution of which might be possible with a fully operable neutron interferometer.

It is the aim of this chapter to explain the principles of X-ray and neutron interferometry, to review their present state, to present what applications have been possible up to now, and, if possible, to ponder the future development.

In Section 4.1, the essentials of X-ray and neutron interferometry will be discussed. For optimizing interferometry and for an exact interpretation of the results, a quantitative understanding of the mechanism of wave splitting and recombining is indispensable. Section 4.2 therefore discusses the functioning of interferometer components on the grounds of the dynamical diffraction of X-rays and neutrons by perfect crystals. In Section 4.3 the most commonly used type, the so-called triple Laue case (LLL) interferometer, is treated in detail. To link realistically with the experiments, absorption is also included in the treatment, and intensity distributions of the beams emerging from the interferometer have been calculated. Geometrical tolerances and thermal as well as positional stability requirements are discussed. Other kinds of interferometers are described in Section 4.4. Applications of angström range interferometers are presented and discussed in Section 4.5.

Up to now four reviews have been written on the subject (*Bonse* [4.7, 8] and *Hart* [4.9, 10]).

4.1 Essentials of X-Ray and Neutron Interferometry

From the experimental and from the theoretical point of view, X-rays and neutrons have more in common with each other than with electron rays. For X-ray and neutron interferometry to become possible, the discovery of a new technique of handling beams of these radiations, namely of Bragg diffraction optics, was essential.

4.1.1 Relation to Electron Interferometry

We shall not discuss electron interferometry here because it is already very well developed and, furthermore, the experimental situation is quite different from that with X-rays and neutrons. As is well known, electron beam processing components such as lenses and mirrors are at hand in a variety of versions and their performance is to some extent comparable to that of optical lenses and mirrors. It is not surprising therefore, that the feasibility of electron interferometry was demonstrated quite a while ago (*Marton* et al. [4.11–13]). The characteristics and performance of electron interferometers of the biprism type [4.14–16] and of the Michelson type [4.17] were investigated in some detail. Averaged inner potentials of electrons in metals were measured [4.18–20]. A quite spectacular use of an electron interferometer was its application to the direct observation of magnetic flux quantization [4.21–24].

In spite of the ease of its operation and of its successes so far, it appears that the application of electron interferometry will not spread very extensively in the near future. A main reason for this may be that for using the electron as a probe of investigation there are competing so successful methods as the various versions of electron microscopy and electron microanalysis. Also, because of the strong interaction of electrons with matter, the necessity of using thin samples may be a handicap for electron interferometry.

4.1.2 Principles of Interferometry

Interferometry may be considered the art to determine the change of a beam with wave state ψ induced by some interaction $\hat{W}$ by the technique of superimposing the altered state $\hat{W}\psi$ with the original state ψ yielding two states $(\psi + \hat{W}\psi)/\sqrt{2}$ and $(\psi - \hat{W}\psi)/\sqrt{2}$ and then measuring the intensity (modulus) of $(\psi + \hat{W}\psi)/\sqrt{2}$ or $(\psi - \hat{W}\psi)/\sqrt{2}$ (Fig. 4.1a).

Of special interest is the case where $\hat{W}$ is a unitary operator, because then the intensity of the beam $(\hat{W}\psi)^*(\hat{W}\psi) = \psi^*\psi = 1$ remains unaffected by $\hat{W}$ so that nothing about $\hat{W}$ can be learned from a simple measurement of the intensities without and with interaction[1]. In more specific language, an unitary operator $\hat{W}$ implies that there is no absorption.

If $\hat{W}$ is unitary we can write $\hat{W} = \exp(i\hat{\phi})$ with $\hat{\phi}$ Hermitian and hence $\hat{W}^* = \exp(-i\hat{\phi})$. The interferometric measurement then gives

$$I = (\psi + \hat{W}\psi)^*(\psi - \hat{W}\psi)/2 = 1 + \psi^*\cos\hat{\phi}\psi = 1 + \langle\cos\hat{\phi}\rangle. \tag{4.1}$$

Consequently the expectation values of $\hat{\phi}$ and hence the properties of $\hat{W}$ may be determined from the measurement of I.

A typical $\hat{W}$ could be the phase shift caused by a parallel-sided or wedge-shaped sample. $\hat{W}$ could also be the rotation of the neutron spin when the neutron passes through a magnetic field in one of the interfering beams.

4.1.3 Experimental Realization of Interferometry with X-Rays and Thermal Neutrons

From the experimental aspect, an essential feature of interferometry is therefore that the interfering beams contain reasonably long sections, typically at least 1 cm, of sufficient separation (at least 0.5 cm) so that a useful sample can be

---

[1]   For sake of simplicity we have assumed ψ to be normalized, which does not limit the generality of the argument.

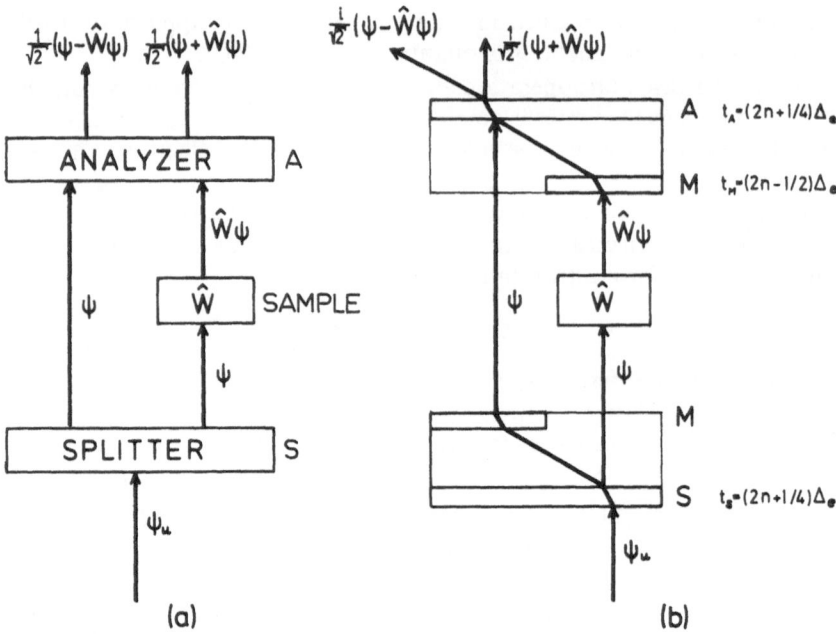

Fig. 4.1. (a) Operator scheme of interferometry. Essential interferometer components are the splitter S, which serves to produce two coherent beams of state ψ, and the analyzer A, which serves for mixing ψ with the altered state $\hat{W}\psi$. $\hat{W}$ is the interaction operator representing the scattering by the sample. Typical mixed states are proportional to $\psi - \hat{W}\psi$ and $\psi + \hat{W}\psi$. The factors $(\sqrt{2})^{-1}$ ensure particle conservation which holds if absorption within the analyzer can be neglected. (b) Realization of an interferometer with Laue case diffraction by four perfect crystals. Besides S and A, mirror crystals M are introduced which help in recombining the interfering beams. Δ_e is the extinction length of the dynamical diffraction theory (4.16). Shown is a special choice of the thickness t_S, t_M, and t_A of S, M, and A, respectively, by which the mixed states become the ones shown. The interferometer is operable also with other values of t_S, t_M, and t_A

placed in only one beam without difficulty. It is practically impossible to establish such a beam geometry with methods borrowed from light interferometry because for X-rays and neutrons no useful conventional lenses and mirrors as beam processing components can be made, because the refractive index of all materials differs from one by no more than 10^{-6} to 10^{-4}. Nevertheless *Maier-Leibnitz* and *Springer* [4.6] were able to operate successfully a neutron interferometer in which beam splitting was accomplished by a biprism. They thus demonstrated that neutron interferometry is feasible in principle, although, because of the very small beam separation of the order of 60 microns, it proved impossible to measure the phase shift with separate samples placed in the interfering beams [4.25].

The essential idea for the realization of an interferometer with widely separated beam paths was to engage Bragg or Laue case diffraction from perfect crystals [4.1, 26–28] for beam handling. Perfection of the crystals is required to

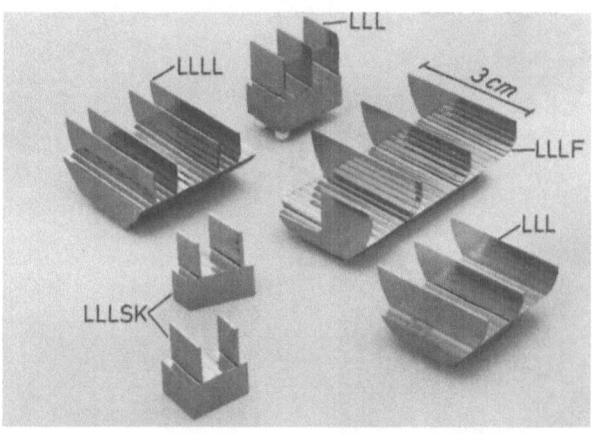

Fig. 4.1c. Examples of actual interferometers, all cut from perfect crystals of silicon. LLL: symmetric triple Laue case interferometers. LLLF: symmetric LLL-interferometer with fore crystal as part of the same single crystal. LLLSK: skewsymmetric LLL-interferometer, two crystal version. LLLL: fourfold Laue case interferometer. LBBL: mixed Laue-Bragg case interferometer; the interferometer can be tuned by tilting the B-parts with respect to the L-parts. For further details see Section 4.4 and also Fig. 4.20

preserve spatial coherence in the handling process. All of the optical instruments which have so far been constructed were manufactured from single crystals of silicon, simply because silicon crystals of suitable quality and size are commercially available at reasonable cost. For the study of defects present in the interferometer crystals themselves, interferometers have also been constructed from germanium and quartz [4.29]. Provided that good quality crystals of other materials can be supplied, there is no reason why the operation of interferometers of materials other than silicon, germanium and quartz should not be possible. Because of usually lower intensity of neutron sources as compared to X-ray sources, in neutron interferometry larger beam cross sections and hence larger crystals are often employed. The quality requirements of the raw single crystals are therefore more severe with neutron interferometry than in the X-ray case.

The manufacture of interferometers of silicon is quite straightforward. The crystals are usually cut with a diamond saw and then chemically polished in a mixture of nitric and hydrofluoric acid. With this technique, proceeding with some care, geometrical tolerances of a few microns can be met.

A disadvantage of the use of Bragg diffraction is that with a given wavelength and d-spacing of the interferometer crystal, the direction of beams in the instrument is no longer freely chooseable. An idea to lift this restriction would be most valuable with respect to the solution of the phase problem of structure determination [4.30, 31], to the problem of X-ray resonators [4.32–36] and X-ray holography [4.37–39], and possibly to Fourier transform spectroscopy [4.10].

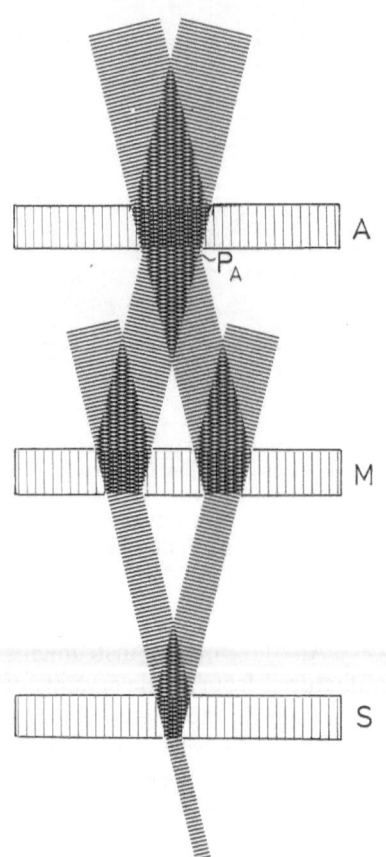

Fig. 4.1d. Standing pattern scheme of a thick LLL-interferometer. The scheme serves as a simple model to explain interference topographs as moiré patterns arising from the superposition of the "grid" of the standing pattern P_A in front of the analyzer A with the "grid" of the Bragg planes of A. For details see Section 4.5.1

4.2 Interferometer Components

The counterparts of the usual components of light optical interferometers like mirrors, beam splitters, etc., in the angström range are Bragg diffracting single crystals, usually named by analogy beam splitter, mirror or analyzer depending on the way they act in the particular case.

We discuss the behavior of these components along the lines of the dynamical theory. For most experimental setups the spherical wave representation of the incident wave is an adequate approach. Following the method first used by *Kato* [4.40, 41], we start from the plane wave diffraction and calculate the spherical wave solutions by a Fourier synthesis of plane wave solutions. For the LLL-interferometer, calculations have been performed for the absorbing [4.42, 43] and the nonabsorbing case [4.44–46]. Performing the calculations for each interferometer component separately, we are able to give rigorous solutions for the various kinds of interferometers by simply combining the component solutions.

4.2.1 Plane Wave Theory of Interferometric Components

Although starting from different equations—Maxwell's equations for X-rays and the Schrödinger equation for neutrons—the solution for a periodic potential requires the same algorithm. We thus write a plane wave for both X-rays and neutrons as a scalar wave in the same notation

$$\mathscr{D}_m^\alpha(r) = D_m^\alpha \exp(2\pi i K_m \cdot r) \tag{4.2}$$

outside the crystal and

$$\mathscr{D}_{mj}^\alpha(r) = D_{mj}^\alpha \exp(2\pi i K_{mj} \cdot r) \tag{4.3}$$

inside the crystal, where D_m^α, D_{mj}^α are amplitudes representing dielectric displacements (X-rays) or wave functions (neutrons) and K_m, K_{mj} wave vectors outside or inside the crystal, respectively. α denotes the polarization state, usually $\alpha = \sigma, \pi$ in case of X-rays, $\alpha = +, -$ in case of neutrons and j numbers the propagation modes, the so-called wavefields, inside the crystal. In the following we will omit α except where special reference to a particular polarization state is made. Real and imaginary parts of complex quantities will be denoted by subscripts r and i, respectively.

The scattering potential of the crystal is described either by the dielectric susceptibility [4.47]

$$\chi(r) = \frac{r_e}{\pi k^2} \cdot \frac{\varrho(r)}{e} \quad \text{(X-rays)} \tag{4.4}$$

with $\varrho(r) = -e\psi^*\psi$ or by the Fermi pseudopotential $V_F(r)$ [4.48] augmented by an absorption term and normalized by the free particle energy $E = h^2 k^2 / 2m_n$

$$\chi(r) \equiv -V(r)/E = -\frac{1}{\pi k^2} \sum_l [(b_c^2 - k^2\sigma_r^2/4)^{1/2} - ik\sigma_r/2]\,\delta(r - r_l) \quad \text{(neutrons)} \tag{4.5}$$

(sum over all lattice points) where the cross section σ_r includes absorption σ_a and incoherent processes σ_s^{inc}. As usual, $-e$ is the charge and r_e the classical radius of the electron, ψ the atomic electron wave function, $k = \lambda^{-1}$ the wave number in vacuum, h is Planck's constant, m_n is the mass of the neutron, and b_c the coherent scattering length.

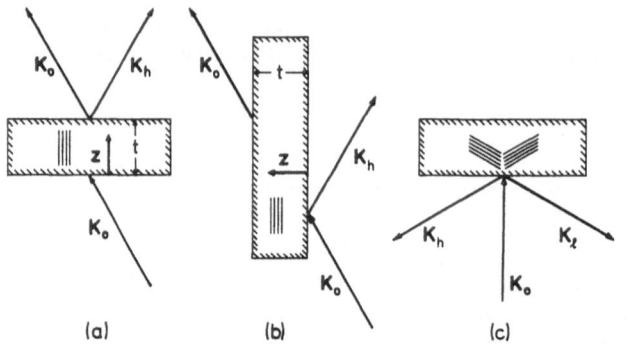

Fig. 4.2. (a) L-beam splitter, (b) B-beam splitter (c) three-beam case beam splitter

Expressions (4.4) and (4.5) can be expanded in a Fourier series summing over all reciprocal lattice vectors g_m. The Fourier coefficients are denoted throughout the paper by χ_m. Absorption is included by complex values of χ and V, respectively, leading to

$$\chi_m = \chi_{rm} + i\chi_{im} \tag{4.6}$$

with χ_{rm}, χ_{im} the Fourier coefficients of $\text{Re}\{\chi\}$ and $\text{Im}\{\chi\}$, respectively. For X-rays in general and for neutrons if $b_c > 0$, we have $\text{Re}\{\chi\} < 0$. Because of the positive sign in the exponent of (4.2) $\text{Im}\{\chi\} > 0$ so that with absorption the intensity of a wave is decreasing in the propagation direction.

In this notation, magnetic neutron scattering is not included but up to now the quality of large magnetic crystals has been too poor to use them as interferometer components.

Beam Splitters

In Fig. 4.2 the different types of beam splitters are sketched. Figure 4.2a shows the Laue case beam splitter for the two-beam case. An incoming wave K_0 generates two wave fields inside the crystal which in turn produce two exit waves, the transmitted (forward diffracted) wave K_0 and the reflected K_h. We define the entrance and the exit surface of a plane parallel crystal slab by $r \cdot z = z_f$ and $r \cdot z = z_f + t$ with z the surface normal and $K_0 \cdot z > 0$.

Matching the boundary conditions

$$\left.\begin{array}{l} \mathscr{D}_0^i(r) = \mathscr{D}_{01}(r) + \mathscr{D}_{02}(r) \\ 0 = \mathscr{D}_{h1}(r) + \mathscr{D}_{h2}(r) \end{array}\right\} r \cdot z = z_f \tag{4.7}$$

$$\left.\begin{array}{l} \mathscr{D}_0^e(r) = \mathscr{D}_{01}(r) + \mathscr{D}_{02}(r) \\ \mathscr{D}_h^e(r) = \mathscr{D}_{h1}(r) + \mathscr{D}_{h2}(r) \end{array}\right\} r \cdot z = z_f + t \tag{4.8}$$

the amplitudes of the exit waves are calculated from the fundamental equations of the dynamical theory [4.45, 49, 50]. Using a bracket notation $\langle m|j|n \rangle$ for the ratio D_n^e/D_m^i of the amplitude of the incoming wave D_m^i and the outgoing wave D_n^e via wave field j, we obtain for the transmitted wave via wave fields 1 and 2:

$$\langle 0|1,2|0\rangle_L = \exp(ia_0 t - iAy)\,[C(y,t) + iyS(y,t)] \tag{4.9}$$

with the abbreviations

$$C(y,t) = \cos[A\sqrt{y^2 + \mathrm{sig}(\gamma_h)v^2}] \tag{4.10}$$

$$S(y,t) = \sin[A\sqrt{y^2 + \mathrm{sig}(\gamma_h)v^2}]/\sqrt{y^2 + \mathrm{sig}(\gamma_h)v^2} \tag{4.11}$$

$$A = \pi t/\Delta_e \tag{4.12}$$

$$v^2 = \chi_h\chi_{\bar{h}}/|\chi_h\chi_{\bar{h}}| \tag{4.13}$$

$$a_m = \pi k\chi_0/\gamma_m. \tag{4.14}$$

γ_h/γ_0 describes the asymmetry of the surface orientation where

$$\gamma_m = s_m \cdot z \qquad m = 0, h \tag{4.15}$$

measures the angle between the surface normal z and the direction $s_m = K_m/|K_m|$ of the wave m. γ_0, γ_h are positive for the Laue case, but γ_h is negative for the Bragg case. Δ_e is known as extinction distance

$$\Delta_e = \sqrt{\gamma_0|\gamma_h|}/(k|C\sqrt{\chi_h\chi_{\bar{h}}}|). \tag{4.16}$$

C is the polarization factor with $C=1$ for neutrons $(\alpha = +, -)$ and normally polarized X-rays $(\alpha = \sigma)$ but $C = \cos(2\Theta_B)$ for X-rays with $\alpha = \pi$. The parameter y measures the deviation from the center of the reflection range including refraction corrections:

$$\begin{aligned}
y &= -[\Delta\Theta_0 \sin 2\Theta_B + \chi_0/2(1 - \gamma_h/\gamma_0)]/(\sqrt{|\gamma_h|/\gamma_0}|C\sqrt{\chi_h\chi_{\bar{h}}}|) \\
&= -k\Delta\Theta_0\Delta_e \sin 2\Theta_B/|\gamma_h| \\
&\quad - k\chi_0\Delta_e[|\gamma_h|^{-1} - \mathrm{sig}(\gamma_h)\gamma_0^{-1}]/2.
\end{aligned} \tag{4.17}$$

Here $\Delta\Theta_0 = \Theta - \Theta_B$ measures the deviation of K_0 from the exact Bragg angle Θ_B calculated from Bragg's equation. In case of absorption, $y = y_r + iy_i$ is complex except for the symmetrical Laue case $(\gamma_0 = \gamma_h)$:

$$\begin{aligned}
y_i &= -k\chi_{i0}\Delta_e[|\gamma_h|^{-1} - \mathrm{sig}(\gamma_h)\gamma_0^{-1}]/2 \\
&= -\mu_0\Delta_e[|\gamma_h|^{-1} - \mathrm{sig}(\gamma_h)\gamma_0^{-1}]/4\pi
\end{aligned} \tag{4.18}$$

with $\mu_0 = 2\pi k\chi_{i0}$, the normal absorption coefficient.

The amplitude ratio for the reflected wave is calculated in a similar way:

$$\langle 0|1, 2|h\rangle_L = i \frac{C}{|C|} \sqrt{\frac{\gamma_0}{\gamma_h}} \frac{\chi_h}{|\sqrt{\chi_h\chi_{\bar{h}}}|} \exp(ia_h t + iAy)S(y, t)$$

$$\cdot \exp[i(a_{rh} - a_{r0})z_f + 2\pi i z_f y_r/\Delta_e]. \tag{4.19}$$

To illustrate the influence of absorption we calculate the intensities of both transmitted and reflected waves:

$$I_T(y_r) = \exp[-\mu_0 t(\gamma_0^{-1} + \gamma_h^{-1})/2]|C(y, t) + iyS(y, t)|^2 I_0, \tag{4.20}$$

$$I_R(y_r) = \gamma_0/\gamma_h|\chi_h/\chi_{\bar{h}}|\exp[-\mu_0 t(\gamma_0^{-1} + \gamma_h^{-1})/2]|S(y, t)|^2 I_0. \tag{4.21}$$

Figure 4.3 shows the results for $t = 10\Delta_e$ and $\mu_0 t = 0, 1, 8$, respectively.

One of the wavefields has anomalously low, the other one anomalously high absorption compared to the normal absorption μ_0 outside the reflection range. Hence the intensities of the exit waves show different features depending on the degree of absorption.

For vanishing absorption ($\mu_0 t = 0$)—with neutron diffraction in silicon, for instance—the exit waves display the well-known Pendellösung oscillations due to slightly different propagation vectors $K_{m1}(y)$ and $K_{m2}(y)$, $m = 0, h$. With $\mu_0 t = 1$, the different excitation strengths of the two wavefields depending on the sign of y_r are clearly visible due to anomalous absorption. For strong absorption, no oscillation occurs but only the anomalously low absorbed wavefield survives near the center of the reflection range, leading to almost equally shared intensities in both 0- and h-wave behind the crystal (Borrmann effect).

The curves of Fig. 4.3 are visible only if the lateral width W of the incident "plane" wave is large compared to the thickness t of the crystal. If on the contrary W is small compared to t and yet the corresponding angular spread small compared to the reflection range, two spatially separated beams are detectable behind the crystal due to different energy flow directions of the two wavefields inside the crystal [4.47]. Hence the Pendellösung oscillations disappear except for that particular y-value where the wavefield beams propagate in parallel directions and remain superimposed with each other.

To obtain the exit positions of the beams for an incident wavepacket of width W in physical space, we represent the incident wave by a distribution $\mathscr{A}(y_r)$ in k-space centered at y_{r0} with a certain width $\pm \Delta y$, $\Delta y \sim W^{-1}$. Then we have to evaluate the following integral

$$D_0^e(r) \sim \int dy_r \mathscr{A}(y_r)\langle 0|1, 2|0\rangle_L \exp[2\pi i K_0(y_r)r]. \tag{4.22}$$

We assume for simplicity $\gamma_0 = \gamma_h = \cos \Theta_B$ and $C = 1$. Introducing a coordinate system (u, v, w) defined in Fig. 4.4 (see also [4.45]) and a coordinate q varying

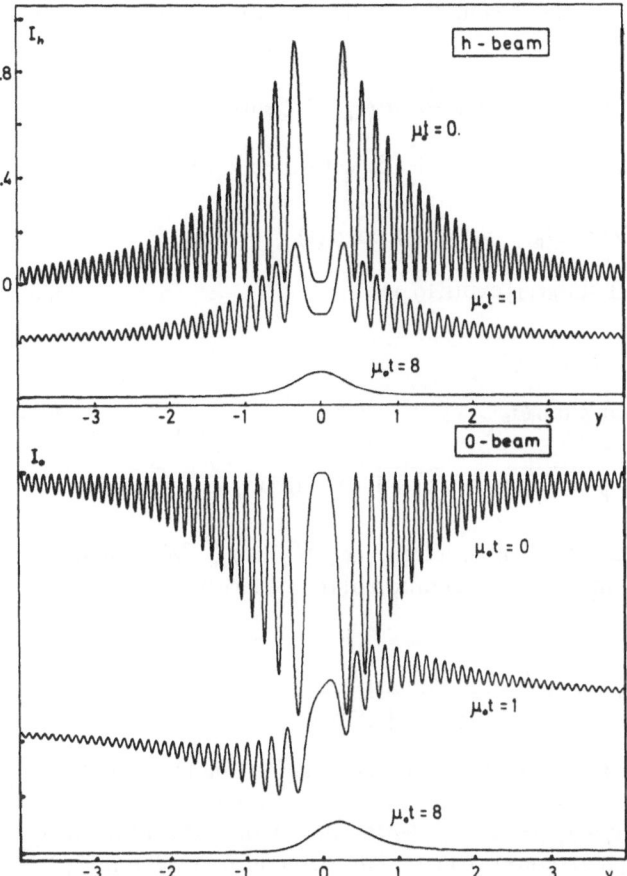

Fig. 4.3. Angular intensity distribution of the beams I_0 and I_h behind an L-beam splitter with varying absorption $\mu_0 t$. Incident intensity normalized to 1. Plane wave case. $y = \pm 1$ at the limits of the range of total reflection in the Bragg case. $t = 10\,\Delta_e$. The rapid oscillations are due to Pendellösungs effects. As is seen, the oscillations fade with increasing absorption. Note that the curves have been shifted vertically with respect to each other

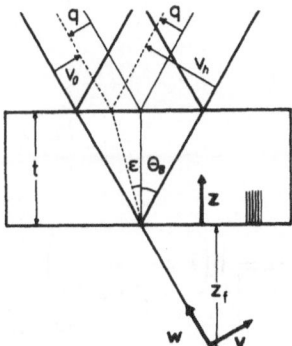

Fig. 4.4. Coordinate system used in the Laue case calculations (see text)

over the cross section of the beam, we calculate with (4.17) for a point $r=(0, v_0, w_0)$:

$$K_0(y_r)r = kw_0 + k\Delta\Theta_0 v_0 = kw_0 + y_r(q - t\sin\Theta_B)/(2\Delta_e\sin\Theta_B). \quad (4.23)$$

If we omit constant phase factors, (4.22) leads with the aid of (4.9, 23) to

$$D_0^e(q_n) \sim \exp[-\mu_0 t/(2\cos\Theta_B)] \int dy_r \mathscr{A}(y_r)[\cos\phi(y)$$

$$+ iy/\sqrt{y^2+1}\,\sin\phi(y)]\exp(iAq_n y) \quad (4.24)$$

with the abbreviations

$$q_n = q/(t\sin\Theta_B) = \tan\varepsilon/\tan\Theta_B \quad (4.25)$$

$$\phi(y) = A\sqrt{y^2+v^2} = A\sqrt{y^2+1} + iA\,\mathrm{Im}\{v^2\}/(2\sqrt{y^2+1}) \quad (4.26)$$

where we have assumed weak absorption, i.e., $|\chi_{rh}| \gg |\chi_{ih}|$. With inversion symmetry ($\chi_h = \chi_{\bar{h}}$) and symmetrical orientation of the surface with respect to the Bragg planes, we find

$$\phi(y) = A\sqrt{y^2+1} + i\,\mathrm{sig}(\chi_{rh})\frac{\mu_0 t|C|}{2\cos\Theta_B}\frac{\chi_{ih}}{\chi_{i0}}\frac{1}{\sqrt{y^2+1}}. \quad (4.27)$$

ε measures the angle of the energy flow direction against the net planes.

Due to the assumptions about $\mathscr{A}(y_r)$, the integral has appreciable values in a certain range $y_{r0} - \Delta y < y_r < y_{r0} + \Delta y$ only, where all functions except the phase factors vary slowly enough to be approximated by their constant value at y_{r0}. We therefore may estimate the integral by the stationary phase method [4.51, 52]. The function $D_0^e(q_n)$ is located around those q_n where the phase is stationary in the interesting y-range. The phase factors involved in (4.24) have the form $\exp[iA(q_n y \pm \sqrt{y^2+1})]$. Hence the q_n are determined by

$$\frac{\partial}{\partial y}(q_n y \pm \sqrt{y^2+1})\bigg|_{y=y_{r0}} = 0 \quad (4.28)$$

leading to two exit points

$$q_{n1,2} = \pm y_{r0}/\sqrt{y_{r0}^2+1} \quad (4.29)$$

where the integral (4.24) has the values

$$D_0^e\left(q_n = \frac{\pm y_{r0}}{\sqrt{y_{r0}^2+1}}\right) \sim \exp\left[-\frac{\mu_0 t}{2\cos\Theta_B}\left(1 \pm \frac{\chi_{ih}}{\chi_{i0}}\frac{1}{\sqrt{y_{r0}^2+1}}\right)\right]\left(1 \pm \frac{y_{r0}}{\sqrt{y_{r0}^2+1}}\right). \quad (4.30)$$

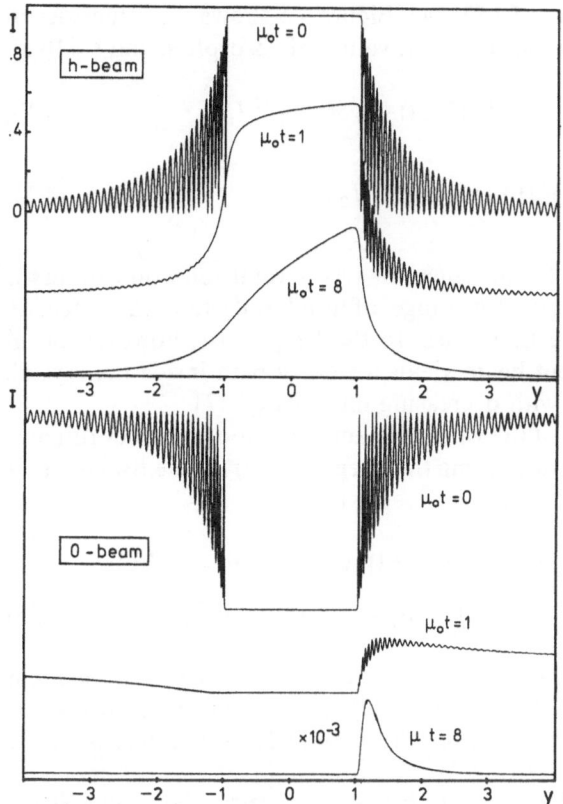

Fig. 4.5. As Fig. 4.3, but Bragg case

Obviously the $(+)$-solution represents the anomalously high and the $(-)$-solution the anomalously low absorbed wavefield.

In most experiments, however, the angular spread W^{-1} of the incident wave is large compared to the reflection range which leads to the spherical wave solution to be investigated in the next subsection.

We proceed with the consideration of the Bragg case beam splitter shown in Fig. 4.2b. A rigorous solution including the back-reflected wavefield from the rear surface of the crystal leads to equations similar to (4.9) and (4.19):

$$\langle 0|1, 2|0\rangle_{\text{B}} = \exp(ia_0 t + iAy)[C(y, t) + iyS(y, t)]^{-1}, \tag{4.31}$$

$$\langle 0|1, 2|h\rangle_{\text{B}} = i\frac{C}{|C|}\sqrt{\frac{\gamma_0}{|\gamma_h|}}\frac{\chi_h}{|\sqrt{\chi_h\chi_{\bar{h}}}|}\exp[i(a_{rh} - a_{r0})z_{\text{f}} - 2\pi iz_{\text{f}}y_{\text{r}}/\Delta_{\text{e}}]$$

$$\cdot S(y, t)[C(y, t) + iyS(y, t)]^{-1}. \tag{4.32}$$

A similar approach was given in [4.53, 54]. Figure 4.5 shows the intensities of transmitted and reflected waves for different values of absorption and $t = 10 \Delta_e$.

$$I_T(y_r) = \exp[-\mu_0 t(1/\gamma_0 - 1/|\gamma_h|)/2]|C(y, t) + iyS(y, t)|^{-2} I_0 , \tag{4.33}$$

$$I_R(y_r) = \frac{\gamma_0}{|\gamma_h|} \left| \frac{\chi_h}{\chi_{\bar{h}}} \right| |S(y, t)|^2 |C(y, t) + iyS(y, t)|^{-2} I_0 . \tag{4.34}$$

In the Bragg case, $\mathrm{sig}(\gamma_h) = -1$ and hence for $|y| < 1$ total reflection occurs if absorption is neglected. Outside the range of total reflection the intensity oscillates [4.55] similar to the Laue case. In the Bragg case, however, beam formation by a narrow incident beam yields a zig-zag path inside the crystal, resulting in many exit beams with decreasing intensity [4.47].

To describe the exit points of the various beams, we choose two coordinates v_0 and v_h defined in Fig. 4.6. In evaluating the integral (4.22) for the Bragg case it is more convenient to transform the expression (4.31):

$$\langle 0|1, 2|0\rangle_B = \exp\{ia_0 t + iA[y - (y^2 - v^2)^{1/2}]\}\{v^2 - [y - (y^2 - v^2)^{1/2}]^2\}$$
$$\cdot \{v^2 - [y - (y^2 - v^2)^{1/2}]^2 \exp[-2iA(y^2 - v^2)^{1/2}]\}^{-1} . \tag{4.35}$$

The notation $(y^2 - v^2)^{1/2}$ is used to define the solution belonging to one Riemann sheet resulting in $|y - (y^2 - v^2)^{1/2}| < 1$ for all y [4.54]. Note that this restriction is not necessary for (4.31, 32) as the functions $S(y, t)$ and $C(y, t)$ are not sensitive to the sign of $(y^2 - v^2)^{1/2}$.

Within the range of total reflection $A(y^2 - v^2)^{1/2}$ gets a large imaginary component for sufficiently thick crystals ($A > 10$) and (4.35) vanishes. For the application of the stationary phase method outside the total reflection range we have to expand the denominator into a power series yielding

$$\langle 0|1, 2|0\rangle_B = \exp(ia_0 t + iAy)(1 - R)\{\exp[-iA(y^2 - v^2)^{1/2}]$$
$$- \sum_{n=1}^{\infty} (-R)^n \exp[-i(2n + 1)(y^2 - v^2)^{1/2}]\} \tag{4.36}$$

with

$$R = [y - (y^2 - v^2)^{1/2}]^2 v^{-2} . \tag{4.37}$$

We simplify the calculation by neglecting absorption and assume $v = 1$, as the beam path is to first order not affected by absorption. From the phase factor $\exp[2\pi i K_0(y)r]$ we extract the y-dependent term $\exp[2\pi i|\gamma_h|v_0 y/(\Delta_e \sin 2\Theta_B)]$ and get for the n-th exit point the equation

$$\frac{\partial}{\partial y}\left[\left(1 + \frac{2|\gamma_h|v_{0n}}{t \sin 2\Theta_B}\right) y - (2n - 1) \mathrm{sig}(y) \sqrt{y^2 - 1}\right]\Bigg|_{y = y_0} = 0 \tag{4.38}$$

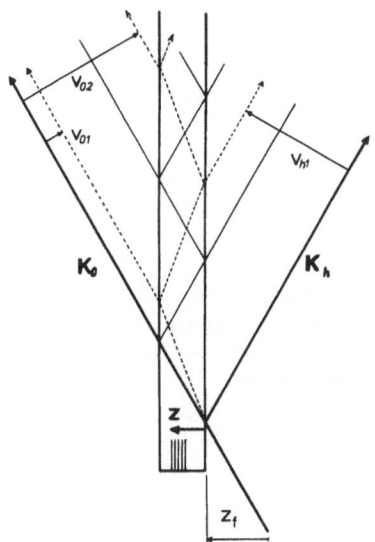

Fig. 4.6. Coordinate system used in the Bragg case calculation (see text)

leading to

$$v_{0n} = \frac{t\sin 2\Theta_B}{2|\gamma_h|}\left[(2n-1)\frac{|y_0|}{\sqrt{y_0^2-1}}-1\right]; \quad |y|>1, \ n=1,2,\dots \tag{4.39}$$

with corresponding intensities:

$$|D_0^e(v_{0n})|^2 = (1-R)^2 R^{2(n-1)}; \quad n=1,2,\dots. \tag{4.40}$$

In a similar way we obtain for the reflected beam:

$$v_{hn} = \frac{t\sin 2\Theta_B}{2\gamma_0} 2n \frac{|y_0|}{\sqrt{y_0^2-1}}; \quad n=0,1,2,\dots, \tag{4.41}$$

$$|D_h^e(v_{h0})|^2 = (\gamma_0/|\gamma_h|)R, \tag{4.42}$$

$$|D_h^e(v_{hn})|^2 = (\gamma_0/|\gamma_h|)R^{2n-1}(1-R)^2; \quad n=1,2,\dots. \tag{4.43}$$

The exit points of the beams in the K_0-direction may have any positive v_0-value with constant spacing between neighboring rays whereas the reflected beams occur at $v_h = 0$ (directly reflected beam) and $t\sin 2\Theta_B/\gamma_0 < v_h < \infty$.

Obviously the Bragg case beam splitter must be operated outside the range of total reflection. With no absorption and $\gamma_0 = |\gamma_h|$ we calculate from (4.39, 42) for the ratio $|D_0^e(v_{01})/D_h^e(v_{h0})| = 1$ the exit point $v_{01} = 1.236\, t\sin\Theta_B$ at $|y_0| = 1.118$.

If on the other hand absorption is not negligible, it is more advantageous to incline the surface against the lattice planes to shorten the length of the path of that beam which is favored by anomalous low absorption.

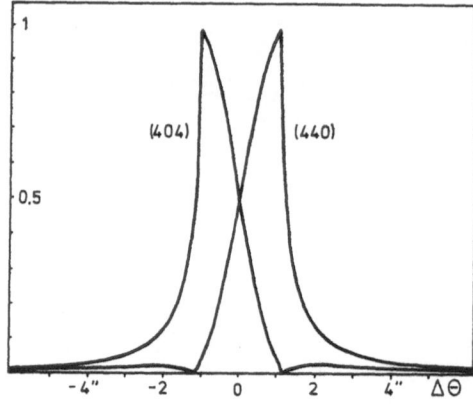

Fig. 4.7. Intensities of the split waves for the silicon (440, 404) three-beam case splitter of Fig. 4.2c as a function of the incident angle of the incoming beam

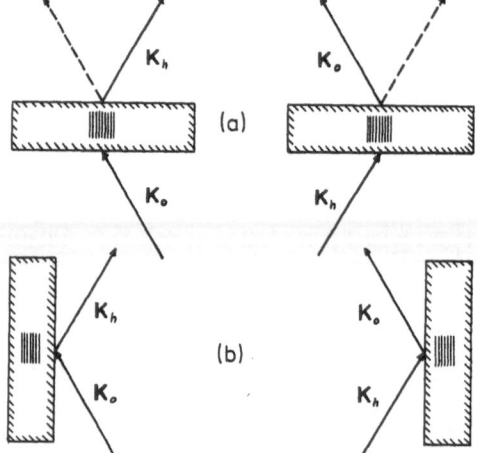

Fig. 4.8a and b. Possible beam mirrors. (a) Laue case, (b) Bragg case

Instead of using together with the directly reflected beam at $v_h = 0$ the transmitted beam $D_0^e(v_{01})$ as a second beam, it is also possible to use the reflected beam $D_h^e(v_{h1})$ which obtains a maximum intensity of $0.148 \, |D_0^i|^2$ at $|y_0| = 1.155$. This arrangement provides two parallel coherent beams with a spacing $v_{h1} = 4t \cos \Theta$ between the maxima. (See also Fig. 4.18a.)

Another kind of beam splitter is the three-beam case Bragg-Bragg splitter (Fig. 4.2c) which has been operated very recently [4.56]. An incoming beam K_0 is simultaneously reflected at two different sets of net planes under Bragg condition, and two beams K_h and K_l emerge from the entrance surface.

The great advantage of this arrangement is the almost complete lack of absorption and of beam widening in the splitting process. On the other hand, only a narrow wavelength band of the order of $\Delta\lambda/\lambda = 10^{-5}$ can simultaneously be diffracted via two reciprocal lattice vectors g_h and g_l. It can be shown that, at least for X-rays, the beam splitter must be operated with all three waves in one plane, otherwise complicated polarization effects will occur resulting in a decrease of interference contrast. For a combination g_h, g_l of net planes the required wavelength for the coplanar case turns out to be

$$\lambda_0 = 2|g_h \times g_l|/(|g_h| \, |g_l| \, |g_h - g_l|) \, . \tag{4.44}$$

For listings of wavelength λ_0 coinciding with characteristic X-ray lines see [4.9, 34]. With the continuous wavelength spectrum of synchrotron radiation or thermal neutrons, a large variety of coplanar three-beam cases is accessible.

The calculation of the amplitudes of a three-beam case is not so straightforward as in the two-beam case because the shape of the dispersion surface varies rapidly with the wavelength. It is therefore far beyond the scope of this chapter to give general amplitude ratios similar to the two-beam case.

Figure 4.7 shows the intensities of the split waves for the (440, 404)-reflection with $\lambda_0 = 1.66276$ Å in silicon [4.56].

Mirrors and Analyzers

For the recombination of the interfering beams, mirrors and analyzers are needed. Any of the previously described components can be used for this purpose. A mirror is simply a beam splitter with just the reflected beam being used (Fig. 4.8).

At first sight the Bragg case with its total reflection range is best suited as a mirror component. However, when combined with a Bragg case beam splitter, the Bragg case mirror is usually operated outside the total reflection range. In connection with a Laue case beam splitter, total reflection can be employed but the different refraction requires some adaptations which are discussed in Section 4.4.

For numerical calculations the list of amplitude ratios given above has to be completed by those factors where the transition occurs from the K_h-wave to either K_0- or K_h-exit waves:

$$\langle h|1, 2|h \rangle_{\mathrm{L}} = \exp(\mathrm{i}a_h t + \mathrm{i}Ay)[C(y, t) - \mathrm{i}yS(y, t)] \, , \tag{4.45}$$

$$\langle h|1, 2|0 \rangle_{\mathrm{L}} = \mathrm{i}\frac{C}{|C|} \sqrt{\frac{\gamma_h}{\gamma_0}} \frac{\chi_{\bar{h}}}{|\sqrt{\chi_h \chi_{\bar{h}}}|} \exp(\mathrm{i}a_0 t - \mathrm{i}Ay)S(y, t)$$
$$\cdot \exp[\mathrm{i}(a_{\mathrm{r}0} - a_{\mathrm{r}h})z_\mathrm{f} - 2\pi \mathrm{i}z_\mathrm{f} y_\mathrm{r}/\varDelta_\mathrm{e}] \, , \tag{4.46}$$

$$\langle h|1, 2|h \rangle_{\mathrm{B}} = \exp(-\mathrm{i}a_h t + \mathrm{i}Ay)[C(y, t) + \mathrm{i}yS(y, t)]^{-1} \, , \tag{4.47}$$

$$\langle h|1, 2|0 \rangle_{\mathrm{B}} = \mathrm{i}\frac{C}{|C|} \sqrt{\frac{|\gamma_h|}{\gamma_0}} \frac{\chi_{\bar{h}}}{|\sqrt{\chi_h \chi_{\bar{h}}}|} S(y, t)[C(y, t) + \mathrm{i}yS(y, t)]^{-1} \, ,$$
$$\cdot \exp[\mathrm{i}(a_{\mathrm{r}0} - a_{\mathrm{r}h})z_\mathrm{f} + 2\pi \mathrm{i}z_\mathrm{f} y_\mathrm{r}/\varDelta_\mathrm{e}] \, . \tag{4.48}$$

(In the Bragg case $rz = z_f$ still denotes the entrance surface with the convention $K_0 z > 0$.)

The partial beams form in the overlap region a pattern with the same periodicity as the lattice planes—at least with an ideally perfect crystal—which can be investigated only by "comparing" it with a detector of that periodicity. A suitable beam analyzer is therefore a reversed beam splitter which makes the divergent beams parallel again, thus transforming atomic scale phase patterns into macroscopic intensity patterns. The sensitivity of the intensity I to a phase difference ϕ between the two interfering beams with intensities I_1 and I_2 is usually measured by the contrast function

$$\gamma = (I_{max} - I_{min})/(I_{max} + I_{min}) \tag{4.49}$$

according to

$$(dI/d\phi)_{max} = \gamma(I_1 + I_2). \tag{4.50}$$

The condition $I_1 = I_2$ is necessary for maximum contrast $\gamma = 1$. From Figs. 4.3, 4.5, and 4.7 we see that except for the Laue case with thick absorbing crystals the intensity ratio between the split beams varies rapidly with y, i.e., with the angle of incidence. In most cases it is possible to employ pairs of equivalent mirrors which affect the amplitudes of the two interfering beams symmetrically. Hence to ensure $I_1 = I_2$ over the full y-range, the analyzer must have the same thickness as the beam splitter and transmit the beam previously reflected in the splitting process and vice versa. Consequently only one of the beams emerging from the analyzer can usually reach 100% contrast. Solely in the limit of thick absorbing Laue case interferometers both of the exit waves show high interference contrast.

Influence of Component Translations

A displacement u of one of the above discussed components with respect to the origin is described by the transformation $\chi(r) = \chi^0(r - u)$ of the unshifted susceptibility χ^0. The Fourier coefficients χ_m are then related to the corresponding χ_m^0 of the unshifted crystal via

$$\chi_m = \exp(-2\pi i g_m u)\chi_m^0 \tag{4.51}$$

leaving the transmitted wave in general unaffected but adding a phase shift $-2\pi i g_h u$ to the reflected h-wave and $2\pi g_h u$ to the reflected 0-wave. Accidental movements of single components with respect to the rest of the interferometer have to be kept small compared to the spacing of the lattice planes, typically $|u| \ll 1$ Å.

Let the translation u depend on time, for instance harmonically,

$$u(t) = u_0 \sin \omega_q t \tag{4.52}$$

where the angular frequency ω_q is low enough to take the corresponding phonon wave vector q as practically zero. In the limit $q \to 0$, inelastic scattering may be completely neglected for X-rays and the statements given above concerning the phase shifts of transmitted and reflected waves hold for $u = u(t)$ as well.

Not so with thermal neutrons, where the relative energy change of the outgoing waves reflected by a vibrating crystal is larger by a factor of 10^5 compared to the X-ray case. The maximum frequency shift of the reflected neutron waves can be estimated to $\omega_{max} \approx 2\pi g_m u_0 \omega_q$ by time differentiating (4.51). For a neutron wave with a phase velocity $v_{ph} = 10^3$ m s^{-1} and a crystal vibration of $\omega_q = 10^2$ s^{-1}, $|u_0| = 0.1$ µm and $|g_m| = 10^8$ cm^{-1} this results in a beat length of 1.5 cm.

Some implications to real interferometers will be mentioned in Section 4.3.

4.2.2 Nonplanar Waves

In most experiments of X-ray and thermal neutron optics, one has to deal with incident waveforms different from plane waves. With a plane wave, by applying particular boundary conditions, a single point on every sheet of the dispersion surface can be excited. With a wave packet in general a certain range of the dispersion surface is coherently excited. The limiting case of a complete uniform excitation of the whole dispersion surface is usually called the spherical wave case.

With the results of the plane wave treatment it is in principle possible to calculate the diffracted wave packet of any arbitrary incident wave packet given by

$$\mathcal{D}_m(r, t) = \int d^3K \int d\omega \mathcal{A}_m(K, \omega) \exp(2\pi i K r - i\omega t). \tag{4.53}$$

To do this, we have to insert the appropriate transition factor $\langle m|j|n \rangle$ into the integral (4.53).

The spherical wave case, first investigated by *Kato* [4.40, 41, 49, 50] and recently adapted to the LLL-interferometer [4.42, 44, 45], has the great advantage of an analytical expression for $\mathcal{A}(K, \omega)$. The Fourier transform of a spherical wave of constant k and ω_0 is given by

$$\mathcal{A}(K, \omega) = [\pi(K^2 - k^2)]^{-1} \delta(\omega - \omega_0) D_m. \tag{4.54}$$

Defining $K = (K_u, K_v, K_w)$ with respect to the coordinate system u, v, w of Fig. 4.4, we carry out the integration over ω, K_w and, restricting $r = (u, v, w)$ to the observation plane ($u = 0$), over K_u [4.45]:

$$\mathcal{D}_n(r) = \exp(i\pi/4 + 2\pi i g_h r \delta_{nh})(kw)^{-1/2} \int dK_v \langle m|j|n \rangle D_m \exp(2\pi i K r) \tag{4.55}$$

where $K = (0, K_v, \sqrt{k^2 - K_v^2})$ and δ_{nh} is the Kronecker symbol. For a constant source distance w we obtain the amplitude $\mathcal{D}_n(r)$ as the Fourier transform of

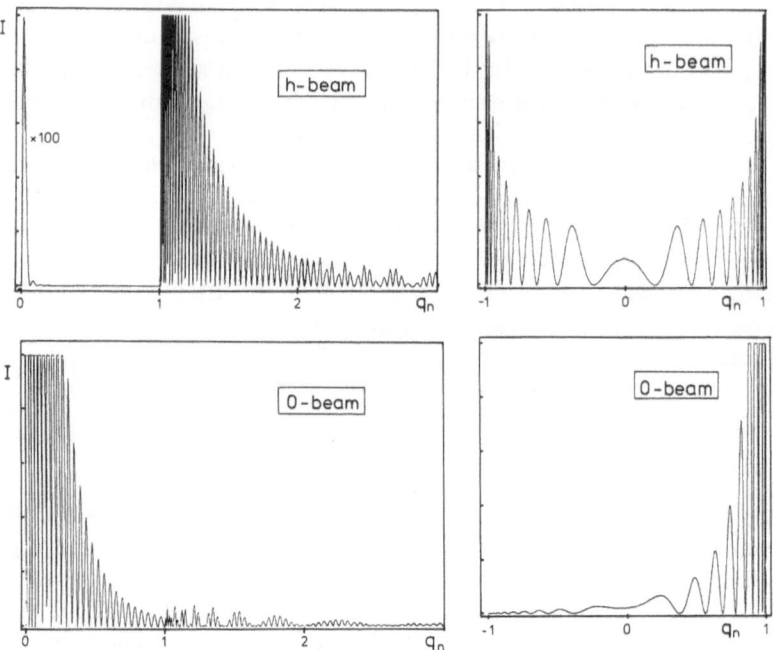

Fig. 4.9. *Spatial* intensity distributions of the beams behind an L-splitter (right) and of the split beams of a B-splitter (left) calculated from the spherical wave theory. The space coordinate q_n is defined by (4.57) together with Fig. 4.4 and (4.58) together with Fig. 4.6, respectively. $t = 10\Delta_e$, zero absorption

$\langle m|j|n \rangle D_m \exp(2\pi i \sqrt{k^2 - K_v^2} w)$ with respect to $K_v = -k\Delta\Theta$. The factor $\exp(2\pi i \sqrt{k^2 - K_v^2} w)$ takes into account the curvature of the asymptotes of the dispersion surface and may be regarded as constant provided $w \ll (k\Delta\Theta^2)^{-1}$ (i.e., typically $\ll 1$ m).

So far we have considered interferometer components exclusively. To calculate the intensity distribution of an incident spherical wave behind the complete interferometer, one can proceed along two different ways. One is to Fourier transform the product of the transition factors of the various components. The other is to convolute the Fourier transforms of the single components according to the convolution theorem:

$$\int dx' f_1(x') f_2(x - x') = 2\pi \int dk F_1(k) F_2(k) \exp(ikx) \tag{4.56}$$

with $F_j(k) = (2\pi)^{-1} \int dx' f_j(x') \exp(-ikx')$. The first method is sometimes easier for numerical calculations whereas the latter gives a better understanding of the main features of the final amplitude and intensity distribution.

The Fourier transforms of the transition factors $\langle 0|1, 2|0 \rangle$ and $\langle 0|1, 2|h \rangle$ for Laue and Bragg geometry were first given by *Kato* [4.49] and *Saka* et al. [4.54], respectively. With the coordinates v_0, v_h illustrated in Figs. 4.4 and 4.6 we define

normalized parameters q_n varying over the cross sections of the beams
a) for the Laue case:

$$q_n = \begin{cases} 1 - 2\gamma_h v_0/(t \sin 2\Theta_B) & \text{0-beam} \\ 2\gamma_0 v_h/(t \sin 2\Theta_B) - 1 & \text{h-beam} \end{cases} \quad |q_n| \leq 1, \tag{4.57}$$

b) for the Bragg case:

$$q_n = \begin{cases} |\gamma_h| v_0/(t \sin 2\Theta_B) & \text{0-beam} \\ \gamma_0 v_h/(t \sin 2\Theta_B) & \text{h-beam} \end{cases} \quad q_n \geq 0. \tag{4.58}$$

With $\theta(x) = 1$ $(x \geq 0)$ and 0 $(x < 0)$ and neglecting common constants the transforms are given below.

$$\langle 0|1, 2|0 \rangle_L \xrightarrow{F} \exp\{-\mu_0 t/4[(1/\gamma_0 + 1/\gamma_h) + q_n(1/\gamma_0 - 1/\gamma_h)]\}$$

$$\cdot [(1+q_n)/(1-q_n)]^{1/2} J_1\left(\frac{C}{|C|} Av \sqrt{1-q_n^2}\right) \theta(1-q_n^2) \tag{4.59}$$

$$\langle 0|1, 2|h \rangle_L \xrightarrow{F} i \sqrt{\gamma_0/\gamma_h} \exp\{-\mu_0 t/4[(1/\gamma_0 + 1/\gamma_h) + q_n(1/\gamma_0 - 1/\gamma_h)]\}$$

$$\cdot J_0\left(\frac{C}{|C|} Av \sqrt{1-q_n^2}\right) \theta(1-q_n^2), \tag{4.60}$$

$$\langle 0|1, 2|0 \rangle_B \xrightarrow{F} \exp\{-\mu_0 t/2[1/\gamma_0 + q_n(1/\gamma_0 + 1/|\gamma_h|)]\} \sum_{n=0}^{\infty} \theta(q_n - n)(-1)^n$$

$$\cdot \left(\frac{q_n - n}{q_n + n + 1}\right)^{n-1/2} \left\{ J_{2n-1}[2vA \sqrt{q_n(q_n+1) - n(n+1)}] \right.$$

$$+ 2\frac{q_n - n}{q_n + n + 1} J_{2n+1}[2vA \sqrt{q_n(q_n+1) - n(n+1)}]$$

$$\left. + \left(\frac{q_n - n}{q_n + n + 1}\right)^2 J_{2n+3}[2vA \sqrt{q_n(q_n+1) - n(n+1)}] \right\}, \tag{4.61}$$

$$\langle 0|1, 2|h \rangle_B \xrightarrow{F} i \sqrt{\gamma_0/|\gamma_h|} \exp[-\mu_0 t/2(1/\gamma_0 + 1/|\gamma_h|)q_n][J_0(2vAq_n)$$

$$+ J_2(2vAq_n)]\theta(q_n) + i \sqrt{\gamma_0/|\gamma_h|} \exp\{-\mu_0 t/2[1/\gamma_0$$

$$+ (1/\gamma_0 + 1/|\gamma_h|)q_n]\} \sum_{n=1}^{\infty} \theta(q_n - n)(-1)^n \left(\frac{q_n - n}{q_n + n}\right)^{n-1}$$

$$\cdot \left[J_{2(n-1)}\left(2\sqrt{\frac{|\gamma_h|}{\gamma_0}} vA \sqrt{q_n^2 - n^2}\right) + 2\frac{q_n - n}{q_n + n} J_{2n}\left(2\sqrt{\frac{|\gamma_h|}{\gamma_0}} vA \sqrt{q_n^2 - n^2}\right) \right.$$

$$\left. + \left(\frac{q_n - n}{q_n + n}\right)^2 J_{2(n+1)}\left(2\sqrt{\frac{|\gamma_h|}{\gamma_0}} vA \sqrt{q_n^2 - n^2}\right) \right]. \tag{4.62}$$

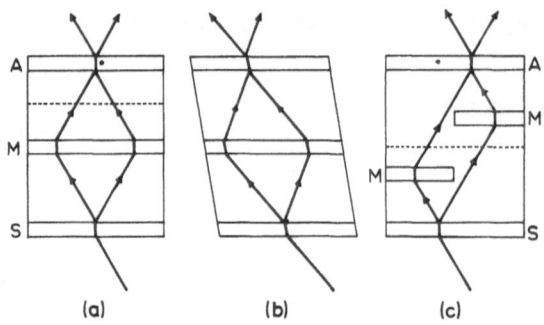

Fig. 4.10a–c. Ray paths in different LLL-interferometers (a) symmetric, (b) asymmetric, (c) skewsymmetric

Figure 4.9 shows the spatial intensity distributions corresponding to (4.59–62) for $t = 10\Delta_e$, zero absorption and the symmetrical case. As is seen, in the Laue case the spatial distribution in the h-beam is symmetrical about the center of the Borrmann fan. Towards the edges the oscillations increase both in amplitude and spatial frequency. The Laue case 0-beam is highly asymmetric. The intensity increases towards that edge of the fan which is the continuation of the incident beam. The Laue case distributions have been experimentally verified in numerous section topographs [4.49].

In the Bragg case, the sharp peak of the h-beam at $q_n = 0$ represents the well-known "surface reflected" beam, commonly referred to as the Bragg reflected wave. The oscillations beyond $q_n > 1$ are very weak and difficult to observe except with very low absorption. Similarly the total Bragg 0-beam is quite weak. However its occurrence is certainly verified by the successful use of Bragg beam splitters [4.27].

4.3 The LLL-Interferometer

Among various interferometer types developed over the last twelve years the triple Laue case (LLL)-interferometer (Fig. 4.10) excels by its broad spectrum of applications whereas the other types remained in the stage of instrument testing. Due to its fairly easy manufacture and straightforward operation in comparison with other types, it marks the beginning of X-ray interferometry [4.1, 26] and also of neutron interferometry [4.2, 57], the possibility of which had been discussed for quite a while [4.6, 27, 58]. With X-rays, absorption usually plays a predominant role in the analyzing process, and a simple optical analogy, the moiré model [4.59], can be applied to interpret the interference pattern. In this limit, coherence properties, geometrical tolerances, and stability requirements have been investigated both theoretically and experimentally [4.43, 46, 60–62]. Neutron interferometry, however, displayed some new properties of the LLL-interferometer because of the absence of absorption requiring a more general theoretical approach, first to interpret the neutron measurements [4.42, 45, 63, 64] and second to link the limiting cases of zero and strong absorption in a uniform representation [4.44].

4.3.1 General Expressions for the Amplitudes

We consider here the LLL-interferometer with arbitrary geometry (Fig. 4.11) and arbitrary absorption but with the restriction to a reciprocal lattice vector $\boldsymbol{g}_h$ common to all components. The influence of relative rotations of single components will be discussed later.

To calculate the amplitudes of the interfering beams, we combine the results of the preceding section and with (4.9, 19, 45, 46) we obtain for an incident plane wave D_0^i the amplitudes
via path I:

$$D_0^{\text{eI}}(y) = \langle 0|1, 2|0\rangle_{\text{L}}^{\text{S}} \langle 0|1, 2|h\rangle_{\text{L}}^{\text{M1}} \langle h|1, 2|0\rangle_{\text{L}}^{\text{A}} D_0^i$$
$$= -v^2 \exp[iQ_0 T_1 + iz_{\text{A}}^{\text{I}}(Q_0 - Q_h)][C(y, t_{\text{S}}) + iyS(y, t_{\text{S}})]$$
$$\cdot S(y, t_{\text{M1}}) S(y, t_{\text{A}}) D_0^i, \tag{4.63}$$

$$D_h^{\text{eI}}(y) = \langle 0|1, 2|0\rangle_{\text{L}}^{\text{S}} \langle 0|1, 2|h\rangle_{\text{L}}^{\text{M1}} \langle h|1, 2|h\rangle_{\text{L}}^{\text{A}} D_0^i$$
$$= i\frac{C}{|C|} \sqrt{\frac{\gamma_0}{\gamma_h}} \frac{\chi_h}{|\sqrt{\chi_h \chi_{\bar{h}}}|} \exp[iQ_h T_1 - i(z_{\text{f}} + z_{\text{S}}^{\text{I}})(Q_0 - Q_h)][C(y, t_{\text{S}})$$
$$+ iyS(y, t_{\text{S}})] S(y, t_{\text{M1}})[C(y, t_{\text{A}}) - iyS(y, t_{\text{A}})] D_0^i; \tag{4.64}$$

via path II:

$$D_0^{\text{eII}}(y) = \langle 0|1, 2|h\rangle_{\text{L}}^{\text{S}} \langle h|1, 2|0\rangle_{\text{L}}^{\text{M2}} \langle 0|1, 2|0\rangle_{\text{L}}^{\text{A}} D_0^i$$
$$= -v^2 \exp[iQ_0 T_2 + iz_{\text{S}}^{\text{II}}(Q_0 - Q_h)] S(y, t_{\text{S}}) S(y, t_{\text{M2}})$$
$$\cdot [C(y, t_{\text{A}}) + iyS(y, t_{\text{A}})] D_0^i, \tag{4.65}$$

$$D_h^{\text{eII}}(y) = \langle 0|1, 2|h\rangle_{\text{L}}^{\text{S}} \langle h|1, 2|0\rangle_{\text{L}}^{\text{M2}} \langle 0|1, 2|h\rangle_{\text{L}}^{\text{A}} D_0^i$$
$$= -v^2 i \frac{C}{|C|} \sqrt{\frac{\gamma_0}{\gamma_h}} \frac{\chi_h}{|\sqrt{\chi_h \chi_{\bar{h}}}|} \exp[iQ_h T_2 - i(z_{\text{f}} + z_{\text{A}}^{\text{II}})(Q_0 - Q_h)]$$
$$\cdot S(y, t_{\text{S}}) S(y, t_{\text{M2}}) S(y, t_{\text{A}}) \tag{4.66}$$

with the abbreviations

$$Q_0 = a_0 - \pi y / \Delta_{\text{e}} \qquad Q_h = a_h + \pi y / \Delta_{\text{e}}, \tag{4.67}$$

$$T_j = t_{\text{S}} + t_{\text{M}j} + t_{\text{A}}. \tag{4.68}$$

Corresponding intensity curves for varying y (intrinsic curves) are given in [4.45].

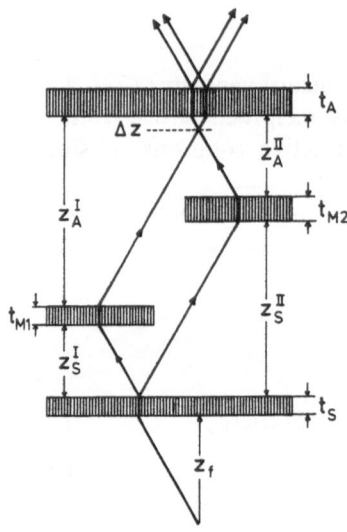

Fig. 4.11. The skewsymmetric LLL-interferometer with arbitrary geometry. The defocus Δz and—with small or zero absorption—the difference $t_A - t_S$ are the essential geometric causes for fringe contrast deterioration

The integrated reflectivity R_m, $m=0, h$, of the interfering beams may depend on a phase ϕ due to a phase object in one path and a spectral distribution $g(k, y)$:

$$R_m(\phi) = \int dk \int dy \, g(k, y) |D_m^{\text{eI}}(k, y) + D_m^{\text{eII}}(k, y) \exp[i\phi(k)]|^2 . \tag{4.69}$$

Note that in the calculation of the integrated reflectivity the phase relation between two different incident plane waves does not enter. For simplicity we assume $g(k, y) = \delta(k - k_0)$ corresponding to a monochromatic spherical wave.

4.3.2 The Ideal Geometry

From the amplitude ratio $D_0^{\text{eI}}/D_0^{\text{eII}}$ one can easily obtain geometrical conditions to ensure $D_0^{\text{eI}}(y) = D_0^{\text{eII}}(y)$ independent of y:

$$t_S = t_A = t , \tag{4.70}$$

$$t_{M1} = t_{M2} ; \quad z_A^{\text{I}} = z_S^{\text{II}} ; \quad z_A^{\text{II}} = z_S^{\text{I}} \tag{4.71}$$

where the last equality was obtained from

$$z_S^{\text{I}} + t_{M1} + z_A^{\text{I}} = z_S^{\text{II}} + t_{M2} + z_A^{\text{II}} . \tag{4.72}$$

Condition (4.71) was first formulated in [4.43] assuming sufficiently thick absorbing crystals. Equation (4.70) is relevant with zero or medium absorption only, i.e., especially in neutron interferometry [4.44, 45].

Equations (4.70) and (4.71) are called the conditions for ideal geometry. Provided (4.70) and (4.71) hold, we obtain for the h-waves the amplitude ratio

$$D_h^{\text{eI}}(y)/D_h^{\text{eII}}(y) = 1 - [vS(y, t)]^{-2} . \tag{4.73}$$

For low $\mu_0 t$-values the relative phase of the interfering h-waves depends on y whereas for $\mu_0 t \gg 1$ the second term in (4.73) vanishes and the two h-waves are in phase, too [4.43]. It can be shown that for small $\mu_0 t$ it is principally impossible to find a geometrical condition to yield $D_h^{\mathrm{eI}}(y) = D_h^{\mathrm{eII}}(y)$ for arbitrary y-values [4.45].

Without additional phase shift on either path the integral intensity of the 0-beam has a maximum, and consequently with zero absorption the h-beam intensity has a minimum. A phase shifting object with $\phi = (2n + 1)\pi$ switches the intensity from the 0-beam to the h-beam. Hence the total intensities of 0- and h-beam oscillate out of phase. The situation differs for sufficiently high absorption; here both 0- and h-partial waves are in phase and so are the oscillations of the total intensity. Only the wavefields with low absorption can reach the exit surface of each crystal wafer leading to an intensity pattern P_A in front of the analyzer with the periodicity of the net planes. With phase shift $\phi = 2n\pi$, the nodes of the pattern point onto the planes of the atomic sites in the analyzer, i.e., on the planes of maximum absorption. Hence only the wavefield with low absorption is generated and the exit beams have maximum intensity (Fig. 4.1d). A phase shift $\phi = (2n + 1)\pi$ moves the antinodes to the atomic sites and the wavefield with high absorption is generated [4.43]. The mechanism described is in effect very similar to the moiré technique in optics where two grids of equal (or nearly equal) spacing are superimposed.

With (4.69) we calculate the contrast function $\gamma(m)$ of the m-beam defined in (4.49). With ideal geometry $\gamma(0) = 1$ for all values of $\mu_0 t$. The interference contrast $\gamma(h)$ of the h-beam, however, shows a complicated behavior with varying thickness and absorption. In Fig. 4.12 $\gamma(h)$ is plotted versus t/Δ_e for various values of $\mu_0 \Delta_e$ assuming equal thickness t for all components. With zero absorption $\gamma(h)$ oscillates about a mean value $\overline{\gamma(h)} \simeq 0.43$. Using the integrated reflectivities averaged over the Pendellösung oscillations [4.44], one calculates $\gamma(h) = 0.56$. With $\mu_0 \Delta_e = 0.045^2$ the mean value $\overline{\gamma(h)}$ drops down to 0.15 for $\mu_0 t \simeq 0.36$ and increases for higher values. With $\mu_0 \Delta_e = 0.15$ the minimum occurs at the same value $\mu_0 t \simeq 0.36$, and $\gamma(h) = 0.9$ is reached with $\mu_0 t = 0.85$. In Fig. 4.3 the Pendellösung oscillations are still present with $\mu_0 t = 1$. Hence it is not necessary to use Borrmann transmission for high interference contrast in the h-beam.

Note that with $\mu_0^{-1} = t_a$ we may write $(\mu_0 \Delta_e)^{-1} = t_a/\Delta_e$. High values of $\gamma(h)$ are obtained for $t/\Delta_e \gtrsim t_a/\Delta_e$.

If one calculates the averaged interference contrast $\gamma(h)$ via the averaged intensities [4.44], $\gamma(h)$ drops to zero at $|\mathrm{Im}\{vA\}| = 0.5$ which corresponds to $\mu_0 t = 0.8$. Hence the averaging over the Pendellösung oscillations appears a somewhat dubious procedure, certainly when one calculates the interference of two beams with rapidly varying relative phases of their plane wave components as it is the case with h-beams.

---

[2] For instance, with Si 220 reflection we calculate for Mo K_α $\mu_0\Delta_e = 0.05$ and for Cu K_α $\mu_0\Delta_e = 0.2$.

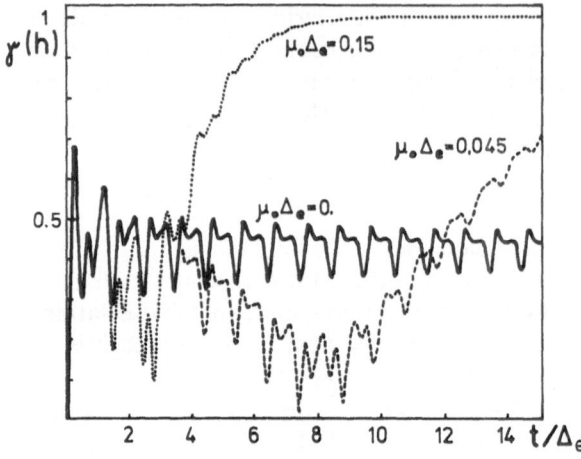

Fig. 4.12. Contrast γ of the h-beam of the LLL-interferometer with ideal geometry for varying absorption μ_0 and thickness $t = t_S = t_M = t_A$ (see text)

4.3.3 Energy Flow Considerations

The results derived so far have been obtained with the plane wave solution solely. The interpretation of the spatial intensity distribution of the exit beams when the interferometer is illuminated through a narrow slit, or the discussion of the imaging capability for inhomogeneous phase objects, however, requires a more detailed investigation of the energy flow of wavefields inside the crystal components.

For the exact calculation of the intensity profiles (4.63–66) have to be Fourier transformed which can be done for an arbitrary geometry on a computer within a reasonable time [4.45]. Assuming $t_S = t_M = t_A$ an analytical solution can be found [4.44].

The essential features of the profiles, however, can be understood by following the ray traces of the single plane wave components through the interferometer. For simplicity we assume symmetrical Laue case and mirrors M of equal thickness t_M.

Usually the divergence of the incident beam is sufficient to excite the whole Borrmann fan so that with low $\mu_0 t_S$ the entrance surface of the mirror M is illuminated over a range $2 t_S \tan \Theta_B$. After three plates, the outgoing beams emerge from an area with width $2 T \tan \Theta_B$, $T = t_S + t_M + t_A$ (Fig. 4.13).

At every entrance surface one single ray component with parameter y is split into two wavefields 1 and 2 with energy flow directions including the angle 2ε (symmetrical about the net planes). Consequently, we have to distinguish between rays which travel through the interferometer via the same wavefield type and those which change their propagation mode. The latter contribute to a certain inner part of the exit area whereas the former sweep over the full range. With increasing absorption, first the rays travelling solely via wavefield 1, the mode with anomalously high absorption, are eliminated, followed by those which travelled twice via type 1, and so on.

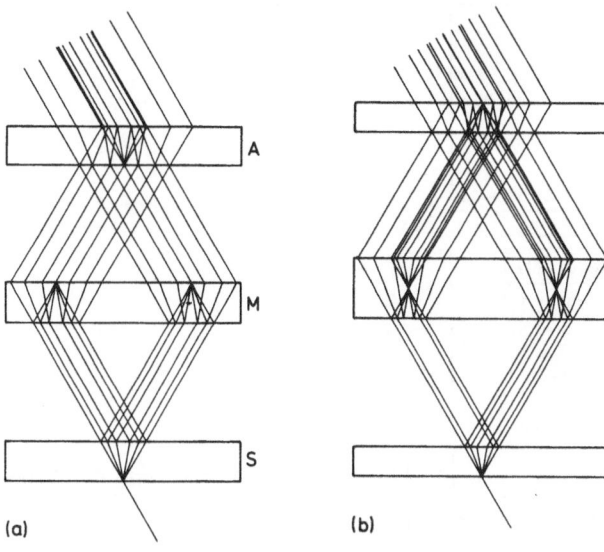

Fig. 4.13a and b. Ray traces through the LLL-interferometer display the side-switching occurring over the full length of the beam path. Note focussing points. (a) $t_S = t_M = t_A$, (b) $2t_S = t_M = 2t_A$

(a)   (b)

Hence with sufficiently high values of $\mu_0 t$ only rays travelling via type 2 can reach the exit surface.

The change of propagation mode causes a side mixing of phases within the inner part of the exit beams which is very troublesome in neutron phase topography (Sec. 4.5) because the phase information is blurred over a range $2 t_A \tan \Theta_B$.

Next we consider two special cases of ideal geometry for zero absorption, namely $t_S = t_M = t_A = t$ and $2t_S = t_M = 2t_A = 2t$ (Fig. 4.13a, b). In the first case a focus F occurs at the exit surface of the mirror M so that we may interpret the intensity profile shown in Fig. 4.14a as the superposition of Pendellösung oscillations of crystal plates with thickness t and $3t$. Since the number of rays contributing to the inner part $2t \tan \Theta_B$ is three times larger than those spreading over the full width $6t \tan \Theta_B$, the intensity ratio between inner and outer parts is roughly 9:1.

The special feature of the second geometry is the focal point at the exit surface, as shown in Fig. 4.14b. Another focus lies in the center of the mirror M so that the intensity profile can be considered as a superposition of Borrmann fans of crystal plates with thickness $4t$, $2t$ and zero.

It is worthwhile to mention that all the features cited above are independent of wavelength. With ideal geometry the LLL-interferometer is achromatic, i.e., the interference contrast does not change with wavelength as long as the change of absorption and dispersion in the sample may be neglected.

4.3.4 Geometrical Tolerances, Influence of Defocusing

Owing to the limited accuracy of the manufacturing process, we always have deviations from the ideal geometry. One can distinguish between several types of

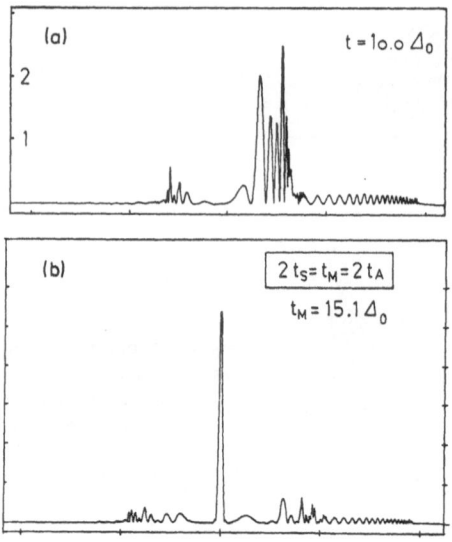

Fig. 4.14a and b. Spatial intensity distribution of the 0-beam of the LLL-interferometers of Fig. 4.13

geometrical aberrations. The most thoroughly investigated deviation is the defocusing Δz (Fig. 4.11). Corresponding rays hit the entrance surface of the analyzer with a separation $2\Delta z \tan \Theta_B$ between them. There are two ways to calculate the resulting interference pattern. One method used in [4.60] is to find those rays which are imaged from a point of the entrance surface of the beam splitter into the same point of the exit surface of the analyzer. By calculating their relative phases one finds a vertical fringe system which modulates the intensity profiles of the outgoing beams. As the fringe position is coupled to the position of the source, the visibility of these fringes is influenced by the source size. With zero absorption the situation is complicated by the superposition of those rays which change their propagation mode while travelling through the interferometer. In this case it is easier to calculate the profiles of path I and II separately with the aid of the spherical wave theory and superimpose these profiles shifted by $2\Delta z \tan \Theta_B$ [4.45]. For thick crystals, averaging over the Pendellösung oscillations yields expressions which allow the determination of the defocus fringe pattern. It turns out that the fringe system already known from the X-ray interferometer is superimposed by another fringe system with one-third spacing which is limited to the central region of the profile [4.44].

Contrast calculations for the 0-beam similar to those of Section 4.3.2 assuming $t = t_S = t_M = t_A = 0.6\,\text{mm}$ and $\Delta z \neq 0$ are shown in Fig. 4.15 (curves 1). For zero absorption (neutrons) the interference contrast drops down to 0.5 at $\Delta z = 11\,\mu\text{m} \cong 0.17\,\Delta_e (t = 9.3\,\Delta_e$, see also [4.44])[3].

With $\mu_0 t = 0.46$ for Mo K_α and Si 220 reflection, we find $\Delta z < 10\,\mu\text{m}$ for $\gamma > 0.5$. For the same reflection with Cu K_α the tolerable defocus increases to

---

[3]   This value agrees with the value $\Delta z = 5 \cdot 10^{-3} T$ with $T \simeq 30\,\Delta_e$ reported in [4.45] but the extrapolation to $T = 180\,\Delta_e$ is invalid.

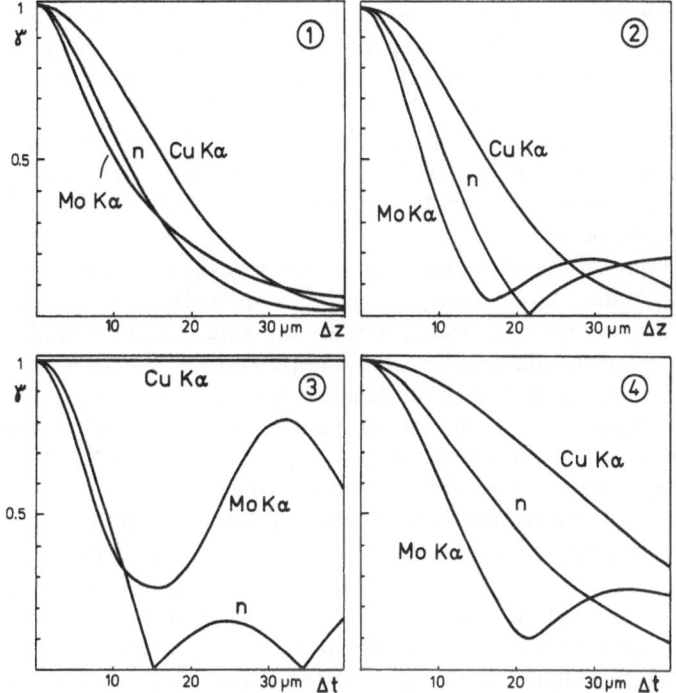

Fig. 4.15. Contrast in the 0-beam of a symmetric LLL-interferometer as function of the various types of geometrical aberrations and for different absorption (see text and Table 4.1)

Table 4.1 Tolerances with various types of geometrical aberrations

Deviation type	Neutrons ($\lambda = 2$ Å)	Mo K_α ($\lambda = 0.71$ Å)	Cu K_α ($\lambda = 1.54$ Å)
1) $\Delta z \neq 0, t_A = t_S$|	11 μm	10 μm	16 μm
2) $\Delta z = \lvert t_A - t_S \rvert$	11 μm	8 μm	16 μm
3) $\Delta t = \lvert t_A - t_S \rvert, \Delta z = 0$	9 μm	8 μm	—
4) $\Delta t = \lvert t_{M1} - t_{M2} \rvert$ $\Delta z = \Delta t/2$	19 μm	13 μm	32 μm
Δ_e	64 μm	35 μm	15 μm
$\mu_0 t$	0	0.46	8

$\Delta z < 16$ μm because of the higher degree of absorption ($\mu_0 t = 8$). *Bonse* and *Te Kaat* found with the oblique analyzer interferometer [4.60] a value of 10 μm with $\mu_0 t \simeq 3.7$. Admissible values of Δz for Si 220 reflection and a wafer thickness of 0.6 mm using different kinds of radiation (thermal neutrons with $\lambda = 2$ Å; Mo K_α, Cu K_α X-ray lines) are listed in Table 4.1.

Another kind of deviation is a defocus Δz caused by a different thickness of beam splitter and analyzer, i.e., $t_A = t_S \pm \Delta z$ (curves 2 in Fig. 4.15 and Table 4.1).

The interference contrast shows for small values of Δz nearly the same behavior, whereas for higher values ($\Delta z \simeq 30\,\mu m$) a partial compensation of the pure defocusing (type 1) occurs. For Cu K_α the shape of $\gamma(\Delta z)$ is exactly the same as with type 1 deviation as the deviation type 3 (curves 3 in Fig. 4.15), $t_S \neq t_A$, $\Delta z = 0$, does not affect the contrast provided the absorption is sufficiently high. But with zero or medium absorption, type 3 deviations turn out to have the steepest decrease of interference contrast with low Δz-values although for $\Delta z \simeq \Delta_e$ appreciable maxima of γ occur.

The last type of deviation calculated in curves 4 assumes a difference between the mirrors of either path, $\Delta t = |t_{M1} - t_{M2}|$, causing a defocus $\Delta z = \Delta t/2$. As expected, the contrast for $\mu_0 t = 8$ is affected only by the pure defocus, but also with low absorption the influence is less severe compared to other types.

Summarizing we may say that with sufficiently high $\mu_0 t$-values only the defocus component (type 1) of any geometrical aberration affects the interference contrast. A typical tolerable defocus is $16\,\mu m$ to ensure $\gamma > 0.5$. With medium absorption this value is only $8\,\mu m$ and with zero absorption we found about $10\,\mu m$. The increase of tolerance for neutrons is mainly due to the larger extinction distance Δ_e (see Table 4.1). If the values given above are expressed in fractions of Δ_e, the tolerances increase with increasing absorption. Roughly speaking the effect of absorption is to make a spherical wave more planar because wave components propagating parallel to the Bragg planes are less absorbed than the rest. Similarly in the case of low absorption, contrast is improved if from outside the waves are made more planar by an asymmetric fore-crystal [4.60] or by using only the center of wave fan with the help of a slit behind the beam splitter [4.65, 66].

4.3.5 Positional Stability, Polylithic Interferometers

As discussed in Section 4.2.1, a relative displacement u of one component causes a phase shift $\pm 2\pi g_h u$ (with mirrors M1 and M2 as a single lamella M the phase shift is even doubled). Hence the positional stability requirements are quite stringent, typically $|u| < 0.2\,\text{Å}$.

The easiest way to guarantee such a high degree of positional stability is to use monolithic interferometers. Furthermore, if the interferometer is manufactured from a material which conducts heat very well and, at the same time, shows little thermal expansion as silicon and germanium do, then a fairly simple thermal shielding will prevent changes due to differential thermal expansion. With X-rays it is sufficient to avoid vibrations of different interferometer components with respect to each other. Consequently for monolithic X-ray interferometers only simple antivibration mounts are needed. Thermal neutrons can suffer phase changes because, owing to their large flight time of typically $5\,\mu s\,cm^{-1}$ ($\lambda = 2\,\text{Å}$), they "experience" different parts of a rigidly vibrating interferometer as shifted. Therefore, in the neutron case fairly good vibration protection is necessary even with monolithic interferometers. There are, however, applications which require the use of polylithic interferometers.

If the reason for the polylithic design is merely the need for more space in the interfering beam paths, for instance in order to facilitate the investigation of large phase shifting objects, then an outlay which has only parallel beams bridging between separated crystal components like the skew-symmetric LLL-interferometer of Fig. 4.10c is very advantageous. It can be shown that movements between the two monolithic subunits indicated in Fig. 4.10c do not cause phase shifts to first order.

In the experimental tests of this polylithic version [4.62], the tolerances of subunit rotations with respect to each other have been investigated. A rotation $\Delta\Theta$ about the Θ-axis (changing the angle of incidence) causes an intensity oscillation due to different inherent translations of the two components of one subunit. The period of this oscillation was found to be $\Delta\Theta = 4.5 \cdot 10^{-3}{}''$. A second rotation $\Delta\varrho$ where the net planes of the subunits are tilted against each other about an axis ϱ lying in the plane of incidence also caused intensity oscillations with a period $\Delta\varrho = 6 \cdot 10^{-2}{}''$. A fringe contrast of 0.5 requires an alignment of $\Delta\Theta = \pm 0.13''$ and $\Delta\varrho = \pm 0.17''$. Once aligned the instrument can be operated without special antivibration mounts.

The situation is much more complicated if the inherent positional sensitivity of polylithic interferometers is a goal in itself, as is the case when the X-ray interferometer is utilized for the precision measurement of lattice parameters. In these experiments the utmost has to be done to ensure mechanical and thermal stability.

For instance, with the two-crystal symmetrical LLL-interferometer [4.61], the ϱ-axis has to be controlled to the order of $10^{-3}{}''$ to avoid rotational moiré fringes which would diminish the interference contrast. To control relative translations of the two subunits, temperature fluctuations have to be kept within $\pm 10^{-3}\,°C$ near the crystals over a 1 h period.

4.4 Other Interferometer Types: Suggestions and Realizations

Soon after the LLL-interferometer was constructed, interferometers combining other beam handling components were proposed and some of them also operated. Besides simply playing variations of a new idea, the intention behind the search for other interferometers is at least twofold: First there are conceivable uses of X-ray and neutron interferometry where the LLL-interferometer is not the best choice of instrument, for instance, the loss of half the intensity in a Laue case mirror is in general undesirable. Furthermore the smearing of beams in a nonabsorbing Laue crystal is usually troublesome in topographic applications, etc.

Secondly, a founded understanding of beam propagation, coherence problems, imaging qualities, etc., is promoted by tackling the same subject from different aspects, i.e., different geometries.

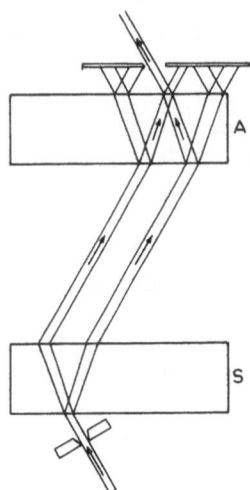

Fig. 4.16. LL-interferometer. Beams which are generated by an incident plane wave are shown. With an incident *spherical* wave both beams cover the full width of the Borrmann fan and no longer stay separated. Interferometry in a limited sense is still possible by putting a sample in half of the beam cross section

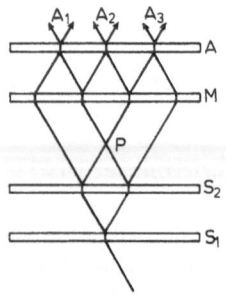

Fig. 17. LLLL-interferometer S_1, S_2: beam splitters, P: sample position

Investigations with interferometers that have surpassed the "instrumental" stage to real applications are treated in Section 4.5.

4.4.1 Laue Case Variations of the LLL-Interferometer

The focusing of an X-ray beam by two successive Laue reflections from two crystal plates of the same thickness as already mentioned in Section 4.3.3 was investigated theoretically [4.67] and experimentally [4.68]. The Fourier transform of $\langle 0|1, 2|h\rangle_L \langle h|1, 2|0\rangle_L$ is essentially a narrow line similar to the focusing of the $t - 2t - t$ LLL-interferometer (Fig. 4.13). For the observation of this focusing effect, however, the values of $\mu_0 t$ must be sufficiently low. With X-rays, fairly thin crystal plates are needed and the widening of the beam within the first crystal is very poor. With neutrons, however, one can use thick crystal plates and thus widen the beam sufficiently to be able to put phase objects into one half of the beam. With the use of an asymmetric fore-crystal, a certain off-center part of the dispersion surface can be exited resulting in two well-separated beams (Fig. 4.16) which do not show Pendellösung oscillations [4.65]. But this LL-

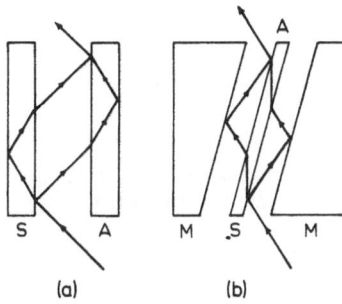

Fig. 4.18. (a) BB-interferometer, (b) BBB-interferometer

interferometer suffers from low intensity because an entrance slit and a beam of low divergence are needed.

In another modification of the LLL-interferometer, a fourth lamella is added resulting in a ray course sketched in Fig. 4.17 [4.69]. All four lamellae have equal thickness t, and the center gap is twice as wide as the outer ones. In this geometry several interferometers are involved superimposing beams in A_1, A_2, and A_3. A calculation of the interfering plane wave components with the transition factors (4.9, 19, 45, 46) yields the result that with low absorption the interference contrast has a maximum in the 0-beams emerging from A_1 and A_3, and in the h-beam from A_2. It is interesting to note that a phase object placed at P influences the interference pattern of all three interfering exit beams. Due to the diffraction focusing already known from the LL-interferometer, the beams crossing at P have a very small lateral width if the incident beam is collimated by a narrow slit. This can probably be used to measure, for instance, phase inhomogeneities of samples by scanning through the point P, because only the differential signal is measured at A_2, whereas the full phase shift can be measured at A_1 and A_3, respectively.

4.4.2 Bragg Case Interferometers

Using Bragg case instead of Laue case components, BB- and BBB-interferometers corresponding to LL- and LLL-interferometers with the same variety of different geometries have been designed.

The two-module symmetric BB-interferometer (Fig. 4.18a) [4.9, 65] utilizes the directly reflected beam $D_h^e(v_{h0})$ and the internally reflected $D_h^e(v_1)$ (see Sec. 4.2.1). Because for sufficient beam separation the beam path inside the crystal must be rather long, this device seems to be more appropriate for neutron interferometry with zero absorption.

The BBB-interferometer (Fig. 4.18b) was experimentally tested [4.27] soon after the first successful operation of the LLL-interferometer. With X-rays the asymmetric geometry is highly recommended to shorten the path length inside the beam splitter and analyzer. The geometrical tolerances were found to be almost equal to those valid for the Laue case interferometers. Due to combined *Bragg reflection* and transmission (Fig. 4.5), the angular range of the exit beams

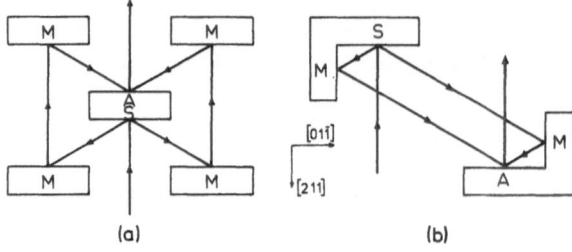

Fig. 4.19a and b. Three-beam case interferometers

is confined to two narrow regions at the edge of the total reflection range. One belongs to anomalously low, the other to anomalously high absorption. Thus in a way similar to the Laue case mirror, half of the intensity is lost with the Bragg case mirror also. All BB- and BBB-interferometers are achromatic.

4.4.3 The Three-Beam Case Interferometer

In the Bragg three-beam case interferometer the transmitted beam of the beam splitter is replaced by a second reflected beam as described in Section 4.2. The geometry required for beam recombination is shown in Fig. 4.19. Although with the layout of Fig. 4.19b one mirror less is needed as compared to Fig. 4.19a, the geometry is lacking a free parameter because the spacings between the components determine the wavelength λ to be focused on the analyzer. On the other hand λ is also fixed by the coplanar three-beam case.

The geometry of Fig. 4.19a has been operated successfully [4.56, 70]. The unique feature of this interferometer type is the complete absence of any beam widening by Borrmann triangles which favors the instrument for high resolution phase contrast topography, for instance. The low absorption losses are advantageous with experiments where high spectral resolution is required. On the other hand the intensity gain is overcompensated by the narrowness of the built-in spectral window of the three-beam case.

4.4.4 Mixed Bragg-Laue Case Interferometers

With the use of both Bragg and Laue case components [4.28] in one interferometer, refraction has to be considered. As long as all crystal surfaces were parallel, a plane wave propagating through the interferometer could be described by the same incidence parameter y in all components.

The influence of refraction with a Laue case splitter and Bragg case mirrors is illustrated in Fig. 4.20a. An incident plane wave 1 is split into waves 1 and 2 (full dots indicating vacuum waves, circles tie-points on the dispersion surface), determined from the continuity of the tangential components of the wave vectors, and then reflected to waves 3 and 4 with tie-points far away from the total reflection range.

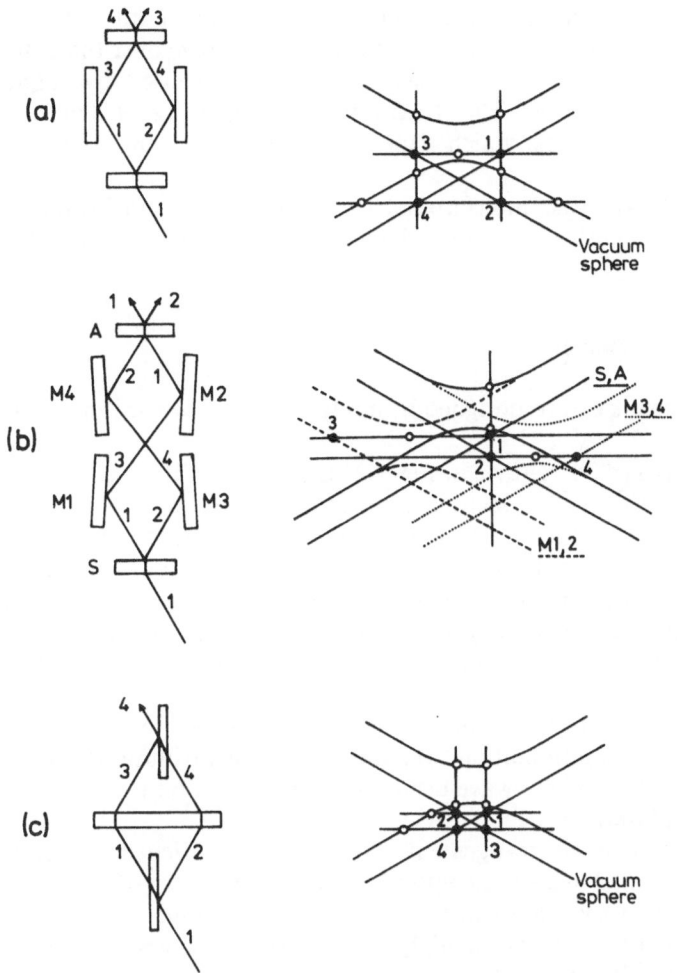

Fig. 4.20. (a) LBL-interferometer. On the right the dispersion surface in K-space is illustrated. Vacuum K-vectors of the beams 1 to 4 are indicated by full dots. Circles represent tie-points of corresponding wavefields inside the crystal. (b) Same as (a), but LBBL-interferometry with Bragg components M1 through M4 rotated to optimize the overall transmission ("tuning"); correspondingly the dispersion surfaces of M1, M2 (dashed) and M3, M4 (dotted) are shifted in K-space. (c) Same as (a) but BLB-interferometer

Obviously the mirror reflection can be much more improved by tuning the Bragg case mirrors to the Laue case reflection range (Fig. 4.20b). Here the Bragg case mirrors are slightly rotated as indicated by shifts of the corresponding dispersion surface. A second mirror reflection is needed to fit the beams 3 and 4 to the Laue case analyzer. A tuning is also possible by heating the Bragg parts (or cooling the Laue parts).

The LBBL-interferometer has been operated successfully both untuned [4.28] and tuned [4.71]. The LBBL version has some advantage over the LBL arrangement because to first order the geometrical error caused by slight asymmetry of surfaces cancels. The BLB-interferometer (Fig. 4.20c) has complicated focusing conditions which depend on wavelength and angle of incidence.

The mixed Bragg-Laue case interferometers all require a monochromatic incident beam as the focusing condition determines the wavelength via the length to width ratio.

4.4.5 Michelson Interferometers

Michelson interferometers can be designed by applying special four-beam cases to beam handling.

The first such design [4.34], shown in Fig. 4.21a, can be operated near room temperature with Ge $(\bar{2}0\bar{2}, 40\bar{4})$ reflections and Co K_{α_1} radiation.

A second design [4.10] (Fig. 4.21b) uses $(\bar{4}00)$ and $(0\bar{4}0)$ reflections in the mirrors and $(\bar{2}20)$ reflections for beam splitting and analyzing. With silicon, this interferometer would work with $\lambda = 2.7154\,\text{Å}$ at room temperature.

With the availability of coherent synchrotron sources the fit to a particular X-ray line is no longer necessary. Neutrons from a reactor have a continuous spectrum anyhow.

With the first type of interferometer, input and output locations are well separated thus avoiding "contamination" of noninterfering contributions which even with a monochromatic plane wave occur via direct reflection by the $(\bar{2}\bar{2}0)$ planes of the beam splitter.

Due to the highly dispersive arrangement of reflections the Michelson X-ray interferometers are confined to a wavelength range $\Delta\lambda/\lambda$ of the order of typically 10^{-5}. Any spectroscopy would have to be done within this range.

It may be noted that because of the highly symmetric four-beam geometry $[(\bar{2}0\bar{2}, 40\bar{4}, 20\bar{6})$ and $(\bar{2}20, \bar{2}\bar{2}0, \bar{4}00)$, respectively] at every surface Bragg reflection, the incident beam is also back-reflected by π. In addition wavefields with energy transport directed into the crystal are generated.

4.5 Applications

The following applications of X-ray and/or neutron interferometry make a useful contribution to other fields: study of defects in nearly perfect crystals, absolute measurement of lattice parameters, precise determination of X-ray scattering factors f and coherent neutron scattering lengths b_c, X-ray and neutron phase contrast microscopy, and measurement of neutron spin rotation and of magnetic scattering of neutrons. Beyond these are some applications like the study of the role of gravity in neutron interference, the phase problem or

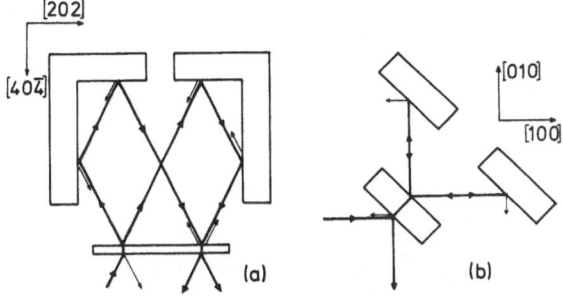

Fig. 4.21a and b. Michelson interferometers, four-beam case. Unwanted reflections are indicated by thin arrows

holography, Fourier transform spectroscopy, resonators for X-ray lasers, etc., which, if realizable, would be of great interest and possible value, but which at present still lack experimental feasibility.

4.5.1 Moiré Technique and Defect Study

The optical moiré interpretation [4.59] explained in Section 4.3 with Fig. 4.1d is very helpful for understanding the imaging of defects with the interferometer. It assumes that a moiré pattern is formed by the superposition of the two "grids" represented by the analyzer A and the standing pattern P_A in front of it. Dislocations or inhomogeneously distributed point defects such as impurities, vacancies, interstitials or clusters of any of these are associated not only with atomic scale displacements (which, because of the limited spatial resolution of the necessary lensless imaging process, remain undetected) but also with long range strain fields. These strain fields represent locally varying lattice dilatations $\Delta d/d$ combined with lattice rotations $\Delta \varrho$, which according to the optical model cause dilatation and rotation moirés with spacings $\Lambda_D = d(\Delta d/d)^{-1}$ and $\Lambda_R = d/\Delta \varrho$, respectively. Similarly, when present in crystals S or M, such strain fields will in effect deform the pattern P_A and become imaged correspondingly.

   With nonabsorbing crystals the result of a grid displacement is switching the intensity among the two outgoing beams 0 and h, primarily due to a translation-induced phase shift of the diffracted amplitude [see (4.51)]. However, such a phase shift can also occur within a single interferometer crystal while the wavefields are transmitted because of tie-point migration [4.72] in a slightly deformed lattice. In practice both effects will be present. With a uniform deformation the tie-point migration is proportional to the total path length within the crystal. Hence the tie-point migration effect becomes more important with thicker crystals. With neutrons, where absorption is mostly negligible, it appears therefore advantageous to use thin wafers S, M, A to obtain a uniform phase across a given beam area.

   Figure 4.22a is the moiré topograph of the interferometer shown in Fig. 4.22b [4.73]. The interferometer was cut from a cylindrical silicon crystal about 8 cm in

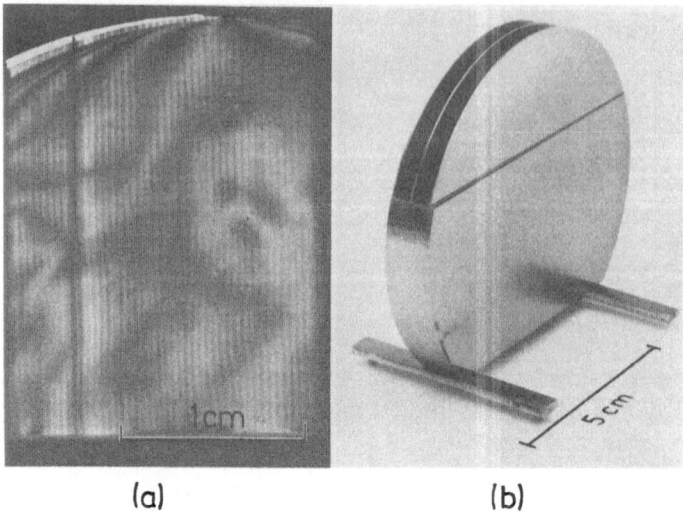

(a)   **(b)**

Fig. 22. (a) Moiré topograph of the interferometer shown in (b). The vertical line structure is the result of stepwise exposure. Note the bending of the vertical lines whenever they enter regions of opposite phase. This effect is assumed to be due to the correlation of the phase shift with tie-point migration of the wavefields in the crystal's deformation fields see text) (after [4.73])

diameter with growth direction $\langle 100 \rangle$ parallel to the axis of the cylinder. The topograph was taken in sections of about 460 µm by shifting the beam between subsequent exposures sidewise by that distance, as may be seen from the vertical line structure. The crystal was free of dislocations. The black and white pattern is caused by strain fields probably due to the inhomogeneous distribution of point defects present in the lattice.

Clearly visible are concentric circles which correlate with the rotation during crystal growth. Furthermore, although the vertical slit defining the exposure sections was straight, it may be seen that the vertical lines are bent or sidewise shifted in correspondence with the black and white phase pattern. The shift is probably caused by tie-point migration. Consequently, from correlation with the moiré pattern, it is concluded that the pattern is rather caused by the presence of inhomogeneous strain within the crystal than by a locally varying translation of the lamellae. Of course, both kinds of strains are not independent from each other. Nevertheless the observation of the bent exposure lines may be the clue for a more complete interpretation of strain moiré patterns.

The correctness of the simple moiré interpretation in cases of homogeneous rotations or dilatations of the analyzer has been demonstrated repeatedly [4.59, 61, 73]. X-ray moirés of single dislocations were first observed by *Bonse* and *Hart* [4.59]. *Hart* [4.74] verified the dependence of the moiré pattern of several dislocations on the choice of Bragg reflection and showed that the Burgers vector can be completely determined with X-ray interferometry. *Christiansen* et al. [4.75] studied an array of 60 dislocations in silicon from their combined moiré pattern. Defects in quartz and germanium interferometers have been examined by *Hart* [4.29]. *Gerward* et al. [4.76, 77] have applied X-ray moiré topography to the study of lateral strains induced by the implantation of 80 keV Ar ions. The thin damaged surface layer resulting from implantation stresses the bulk crystal below, quite similar to the stresses exerted by a thin metal film evaporated on the analyzer [4.78].

The wavelength dependence of moiré fringes was studied by *Eiramdshyan* and *Besirganyan* [4.79] for $Cu\,K_\alpha$, $Ni\,K_\alpha$, $Co\,K_\alpha$ and $Fe\,K_\alpha$. The moiré fringes vary because with different wavelengths different parts of the mirror contribute to the pattern.

Within the limits of the validity of the simple moiré interpretation, X-ray moiré topography is capable of measuring precisely shifts of fractions of angstroms by evaluating fringe positions. It is evident that the sensitivity is limited by the maximum moiré fringe distance Λ_D or Λ_R which is observable. This is usually given by the maximum size of crystal, or, more precisely, the maximum area of the crystal which is homogeneous enough to yield an undisturbed pattern. Consequently with moiré interferometry a gain in strain sensitivity is always coupled with a loss in spatial resolution. The same limitation applies to ordinary moiré topography performed by superimposing two crystal lattices in direct contact [4.78, 80–82].

4.5.2 Absolute Measurement of Lattice Parameter

The spacing d of any moiré generating grids can be measured by shifting one of the grids with respect to the other a known distance X and counting at the same time the number N of resulting moiré fringe passages. Then $d=X/N$.

The importance of this technique when applied to atomic lattices scanned with X-rays or thermal neutrons [4.43] lies in the fact that, unlike conventional lattice parameter measurements, the wavelength of the radiation used does not enter the value of d. Therefore an absolute and direct d-measurement is at hand if only X is measured in standard length units. The latter is feasible by determining X with a light optical interferometer operated directly with the length standard lamp. The light interferometer must be rigidly coupled to the X-ray interferometer (Fig. 4.23a).

Absolute determinations of d become very important once the accuracy exceeds the error of present conventional d-measurements. With a calibrated crystal the following fundamental constants or ratios of fundamental constants can be determined with considerably increased accuracy: Avogadro's constant, h/e, $h/m_e c$, hc/m_n, where h is Planck's constant and m_n the neutron mass.

Furthermore, through the interconnection of most fundamental constants, some constants not mentioned will also have improved accuracy. For the actual experiment a device is needed by which the analyzer A can be translated very smoothly, i.e., with jumps of less than 0.01 Å and over a distance of at least

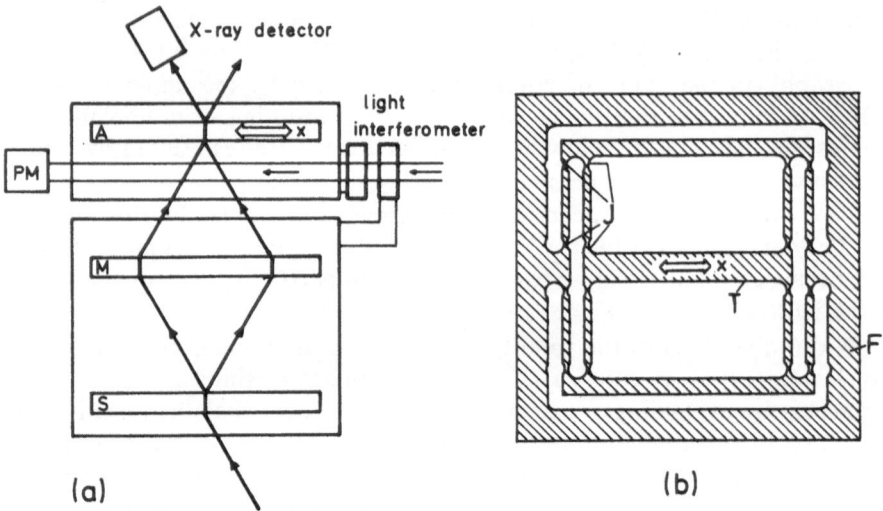

Fig. 4.23. (a) Example of a two-crystal LLL-interferometer coupled with a light optical interferometer. (b) Traverse unit used with (a) for smooth displacements in the angstrom range. The outer frame F remains fixed. The inner part T acts as traverse which is capable of a rotation free translation X through pure elastic deformation of the "joints". Size about 300 mm by 300 mm by 40 mm (after [4.83])

0.1 mm or preferably 1 mm. At the same time, so that the structure of the moiré pattern in the translation process is not changed, rotations about an axis normal to the surface of A should be kept smaller than 0.005 s of arc. The feasibility of making a translation stage with a range of 1 mm and a parasitic rotation of no more than 0.001 s of arc was demonstrated by *Hoffroge* and *Rademacher* [4.83] (Fig. 4.23b).

Of the three experimental groups which took up the actual measurement, two use LLL-interferometers with the analyzer A or the splitter S separately mounted on a special translation stage in which the elastic deformation of spring elements facilitates the necessary smooth translation (*Bonse* and *Te Kaat* [4.61], *Deslattes* [4.84–86]). *Hart* [4.87] uses a monolithic LLL-interferometer which has an elastic spring strip device as part of the same silicon crystal. For the measurement of the displacement X, either a multiple beam interferometer of the Fabry-Perot type with extra optical mirrors [4.88, 89] or a two-beam interferometer with silicon mirrors polished directly onto the interferometer crystal [4.90, 91] is used.

Preliminary results of absolute d-measurements of silicon [4.89, 92–94] indicate an accuracy of 0.15 ppm standard deviation. More work in this direction is certainly necessary. The experiments of the National Physics Laboratory, Teddington, and of the Physikalisch-Technische Bundesanstalt, Braunschweig, are still in progress. The problem of optical and X-ray fringe processing was recently discussed by *Basile* [4.95].

4.5.3 Determination of Scattering Factors f and Coherent Scattering Lengths b_c

The interferometric method is very well suited to measure precisely forward scattering factors f for X-ray and coherent scattering lengths b_c for neutrons. With a precise measurement it may become possible to separate the contributions to f and b_c from the different interactions. Such a separation aids to understanding of the fundamental mechanisms of X-ray and thermal neutron scattering and, possibly, certain intrinsic properties of the neutron itself.

For X-rays we have for the scattering factor of an atom with atomic number Z:

$$f = f_Z + Z^2 m_e/M + f' + \mathrm{i} f''. \tag{4.74}$$

f_Z is the coherent scattering factor of a free atom in the short wavelength limit. The second term is the nuclear Thomson scattering which in view of the increased accuracy cannot be neglected [4.3]. f' is the correction due to anomalous dispersion near absorption edges; f'' accounts for absorption.

For atoms in the solid, a modulation of the photoelectron states by backscattering from neighbouring atoms can occur in the absorption process, resulting in a fine structure of f'' on the short wavelength side of the edge (Kronig structure, EXAFS [4.96]) which can be exploited to get information on the nearest neighbor configuration in the sample crystal. Since f' and f'' are related by a Kramers-Kronig relation [4.97], the fine structure is also present in f'. For the K-edge of Ni this was recently measured with an interferometer operated with synchrotron X-rays [4.98]. The knowledge of exact values of f' is important for a certain method to find the phase in structure determination [4.99].

For neutrons the real part of the coherent scattering length b_c for unpolarized nuclei is given [4.100–102] by

$$b_c = b_{nc} + \gamma r_e f_m(Q)\langle S\rangle \boldsymbol{\sigma} q + \frac{m_e}{2m_n}\left(\gamma + \frac{\varepsilon_2}{e\mu_n^2}\right)\gamma r_e[Z - f(Q)] + \frac{m_n}{\hbar^2}\alpha Z^2 \frac{e^2}{R_c} \tag{4.75}$$

where b_{nc} is the coherent nuclear scattering length, γ the neutron magnetic moment expressed in nuclear magnetons μ_n, $f_m(Q)$ the magnetic and $f(Q)$ the (X-ray) atomic form factor as a function of the momentum transfer Q, $\langle S\rangle$ the mean magnetization, $\boldsymbol{\sigma}$ the spin of the neutron, $q = |Q|^{-1}Q(|Q|^{-1}Qh) - h$ the magnetic interaction vector, h a unit vector in the direction of the magnetization of the ion or the atom, ε_2 the second moment of the intrinsic charge distribution of the neutron, α the electric polarizability of the neutron, and R_c the cutoff radius of the Coulomb field of the atom.

The second term is due to the interaction of a magnetic atom or ion with the field H and if applicable is of the same order as b_{nc}. The third term is due to the Foldy interaction [4.103]; the fourth term is the contribution of an induced electric dipole moment of the neutron [4.104]. The Foldy term is three to four and the induced dipole term roughly eight orders of magnitude smaller than b_{nc}. As is seen, the Foldy contribution is zero for the forward direction and can thus be determined as the difference of an interferometric measurement and an accurate measurement with $Q \neq 0$.

In the experiment the λ-thickness t_λ is measured. With it we calculate

$$f = 2\pi (r_e N \lambda t_\lambda)^{-1} \qquad (4.76)$$

for X-rays and

$$b_c = 2\pi (N \lambda t_\lambda)^{-1} \qquad (4.77)$$

for neutrons. N is the number of atoms per unit volume.

t_λ is measured with an LLL-interferometer. One can use either a parallel-sided sample which is placed in both beams and rotated about a vertical axis (Fig. 4.24a) [4.2, 105] or a pair of equal but oppositely sloped wedges which are then translated in a direction parallel to the Bragg planes (Fig. 4.24b) [4.105]. In either case t_λ is calculated from the intensity oscillations as a function of the sample movement. Photographic fringe recording of a wedge combined with a parallel-sided sample is also employed (Fig. 4.24c) [4.98, 106–108]. *Cusatis* and *Hart* [4.109, 110] measured the fringe shift caused by a parallel-sided sample with the scanning X-ray interferometer (Fig. 4.24d) [4.87]. For an accurate measurement, the refraction of the surrounding air has to be taken into account. Another important factor is exact geometry and homogeneity of the sample. The measurement of the sample thickness can be eliminated if the phase shift is measured at more than one wavelength and if the wavelength dependence, at least in a limiting expression, is known [4.109, 110]. Measurements of f' near absorption edges can best be accomplished by using continuum sources like synchrotron radiation [4.98, 107] or Bremsstrahlung sources [4.109, 110]. Then harmonics $\lambda/2, \lambda/3 \ldots \lambda/n$ from suitable multiple reflection monochromators [4.111] are at hand for the sample thickness elimination.

As follows from (4.76) and (4.77) for the calculation of f and b_c from t_λ, one has also to determine the wavelength λ. With X-rays it is usually not difficult to find a characteristic line or some known absorption structure for calibration.

With neutrons the precise λ-determination is not so straightforward. *Bauspiess* et al. [4.112] therefore operated their neutron interferometer in combination with a specially designed germanium-silicon monochromator [4.113]. It consists of a germanium crystal glued to a silicon crystal at a certain angle β of misorientation in the conventional $(+, -)$ double crystal setting. Because of the difference Δd between the Bragg spacings of the two crystals, only a certain wavelength band $\Delta \lambda$ at a center wavelength λ_0 is transmitted. λ_0 and $\Delta \lambda$ are given by [4.113]:

$$\lambda_0 = [4 d_A d_B / (1 + X^2)]^{1/2} \qquad (4.78)$$

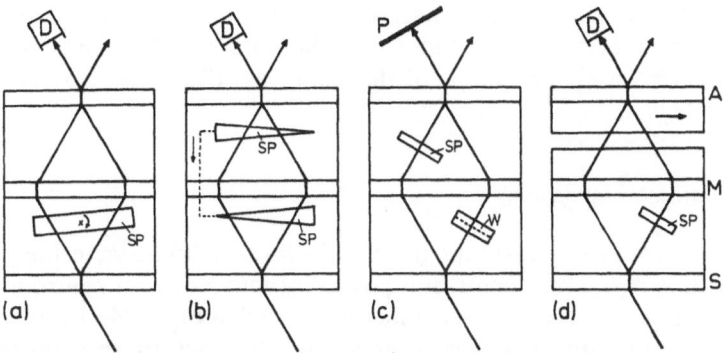

Fig. 4.24. Methods used for f or b_c measurement. SP: Specimen. W: auxiliary wedge for the production of a fringe pattern the shift of which through the specimen is measured (see text)

with the abbreviation

$$X = \beta^{-1}(d_B - d_A)/(d_A d_B)^{1/2} \tag{4.79}$$

and

$$\Delta\lambda/\lambda_0 = \beta^{-1}\Delta\Theta_B \cos^2\Theta_A . \tag{4.80}$$

d_A, d_B are the spacings of the crystals and $\Delta\Theta_B$ is the intrinsic width of the reflection curve according to dynamical theory. With a given pair of crystals, $\Delta\Theta_B$ is to be taken for the material with the largest $\Delta\Theta$-value.

Precise b_c values for Sn, Al, V, Bi, and Nb which were measured (Fig. 4.25a) with that instrumentation are given in [4.100].

Interferometric f- (or $1-n$)-measurements, respectively, have been performed so far for lucite, Be, Si, Al, LiF, NaF at Cu K_α radiation [4.105, 114]; for LiF, NaF, CaF$_2$ at a number of wavelengths between Cu K_α and Ag K_β [4.108, 115]; for Si at Mo K_α, Ag K_α and two equivalent wavelengths from the Bremsspectrum [4.109]; for Ni in the K-edge range with synchrotron radiation [4.98, 106, 107], and Zr with Bremsspectrum and characteristic radiation [4.110]. *Aboyan* et al. [4.116] measured $1-n$ of water, glycerine, and aethyl-alcohol at Cu K_α radiation.

In a combined Pendellösung and interferometry experiment, *Kato* and *Tanemura* [4.3, 117] used a LLL-interferometer to eliminate the sample geometry in the measurement of f_g of silicon for the 111, 220, 333, 440, and 444 reflections. Probable errors as low as 0.05 % were attained.

A diffracting wedge made of a perfect silicon crystal was used as a sample by *Bohlen* [4.118] to study directly the Laue case dispersion surface. With a diffracting sample not only does the value of $|K_{0j}|$, $|K_{hj}|$, $j = 1, 2$, vary according to the shape of the dispersion surface but also the propagation direction of the wavefield so that defocusing occurs which in this case was compensated by using an interferometer with an inclined analyzer [4.60]. Interesting interference fringes between vacuum waves and those from one branch of the dispersion surface could be observed. The shape of the Laue case dispersion surface in general was verified.

4.5.4 Phase Contrast Topography

In addition to absorption contrast topography which is a well-known technique in particular with X-rays, phase contrast topography with X-rays or neutrons is possible if the sample is imaged in the manner illustrated in Fig. 4.24c. Provided that the sample is of equal thickness everywhere, the contrast seen on the topograph is due entirely to spatial variations of phase shifts and/or of absorption which are caused by inhomogeneities of the sample composition. Whether phase contrast topography is a useful technique depends partly on the relative values of the λ-thickness t_λ and the absorption thickness t_a. A necessary condition is $t_\lambda < t_a$ or even $t_\lambda \ll t_a$. Furthermore, the variation of t_λ and t_a between different materials plays an important role. To substantiate the general discussion, we list t_λ and t_a for X-rays and neutrons for some representative wavelengths and materials in Table 4.2.

For X-rays, t_λ is essentially inversely proportional to the density which changes less than a factor of 10 with Z, whereas t_a, apart from the influence of absorption edges, varies with $Z^{-4}\lambda^{-3}$ and gives most of the contrast except with short wavelengths and light elements, where t_λ and hence phase contrast dominates.

Ando and *Hosoya* [4.119] showed that with samples of biological and mineralogical interest, X-ray phase contrast may be useful.

With neutrons, except for some heavily absorbing materials like cadmium, the maximum sample thickness is no longer limited by absorption but rather by the space available in the interferometer. However, another limit is set by dispersion if the wavelength range used is too large, since, as may be seen from (4.76) and (4.77), $t_\lambda \sim \lambda^{-1}$ (with f and b_c to first order independent of λ). Furthermore, inhomogeneities along the path of the beam play an increasing role with thicker samples. This effect can be used, on the other hand, to detect inhomogeneities by the loss of interference contrast caused by them. With neutrons, phase contrast topography might be a useful tool with magnetic samples. However, no work of this kind has yet been published.

4.5.5 Special Applications in Neutron Interferometry

There is a certain group of experiments aimed specifically at the characteristic properties of the neutron as a particle which has spin 1/2, an anomalous magnetic moment, and mass.

To the hypothetical properties of the neutron—namely electric charge and permanent electric moment—rather low upper values have already been

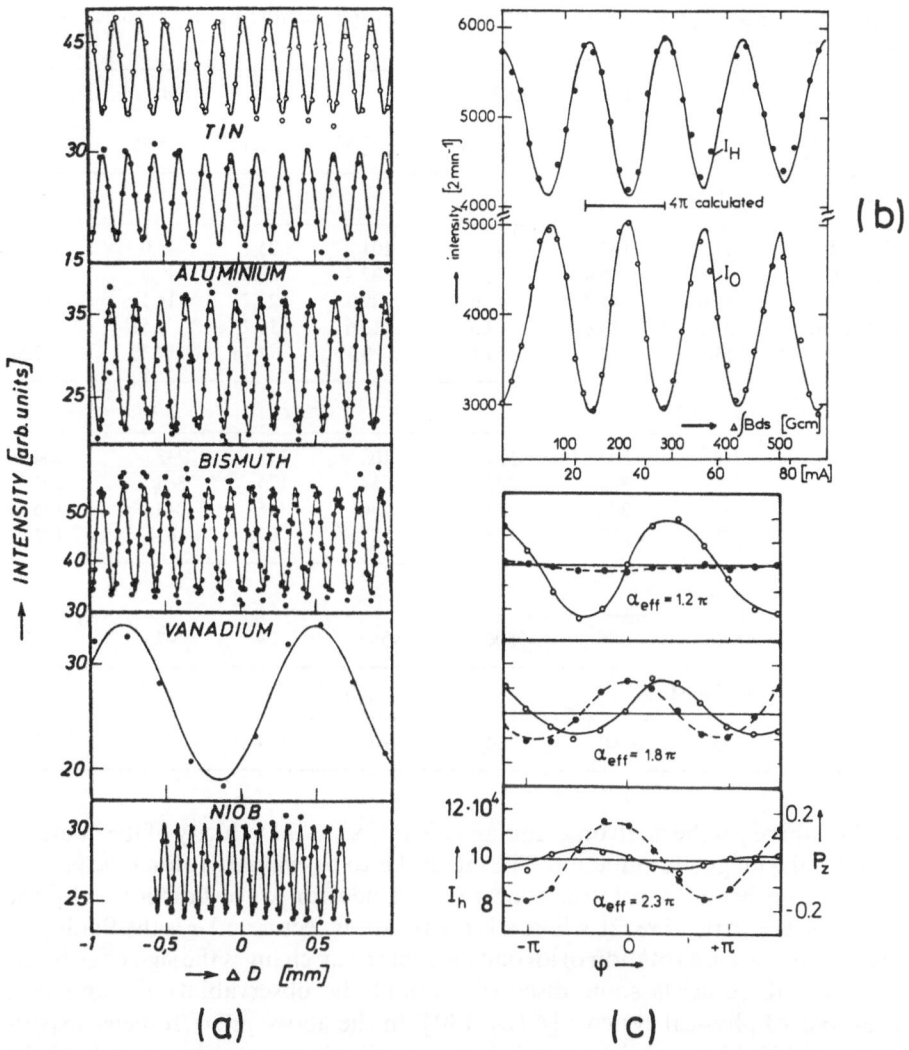

Fig. 4.25. (a) Phase oscillations obtained according to Fig. 4.24a (after [4.100]). (b) Spinor rotation with varying magnetic field (after [4.122]). (c) Interference between nuclear and magnetic phase oscillations (after [4.100])

determined experimentally [4.120, 121] so that at present interferometry appears unlikely to set new limits.

The interaction of the neutron with a magnetic field in one interfering beam was investigated almost simultaneously by *Rauch* et al. [4.122] and *Werner* et al. [4.123]. When passing the magnetic field a rotation of the spinor by

$$\alpha = -\gamma v^{-1} \int \boldsymbol{B} ds \tag{4.81}$$

Table 4.2 λ-and absorption thickness for X-rays ($t_{\lambda x}, t_{ax}$) and neutrons ($t_{\lambda n}, t_{an}$) at some standard wavelengths

	NaCl	Si	LiF	C	Ge	Cd	Pb
ϱ [g cm^{-3}]	2.16	2.33	2.64	3.52	5.32	8.65	11.4
λ [Å]	$t_{\lambda x}$ [μm]						
0.209 (W)	171	153	145	101	75.1	48.0	39.3
0.559 (Ag)	63.5	56.9	54.1	37.4	28.0	17.9	14.7
0.709 (Mo)	50.0	44.7	42.6	29.7	22.2	14.1	11.6
1.541 (Cu)	22.9	20.4	19.5	13.6	10.7	6.51	5.33
2.290 (Cr)	15.3	13.6	13.1	9.15	7.18	4.38	3.58
λ [Å]	$t_{\lambda n}$ [μm]						
0.5	418	600	554	108	348	732	404
1.0	209	300	277	54	174	366	202
1.5	139	200	185	36	116	244	135
2.0	105	150	138	27	87	183	101
λ [Å]	t_{ax} [μm]						
0.709 (Mo)	1130	666	5500	4550	29	42	7.3
λ [Å]	t_{an} [mm]						
1.08	23.1	3340	4	18900	174	0.083	3330

occurs where γ is the neutron gyromagnetic ratio, v is the velocity of the neutron, and $\boldsymbol{B}$ is the magnetic induction. The integral is to be evaluated along the path of the neutron beam. A rotation α acts on the neutron wave function ψ by the operator $\hat{U} = \exp(-i\boldsymbol{\alpha\sigma}/2)$, where $\boldsymbol{\alpha}$ is the rotation vector and $\boldsymbol{\sigma}$ is the Pauli spin vector. As is seen, a rotation of an odd multiple of 2π changes the sign of ψ. In the literature there exists some discussion about the observability of $(2n+1)2\pi$ rotations of physical systems [4.124–130]. In the above interferometer experiments [4.122, 123] and also recently in a molecular beam experiment [4.131], the sign change of ψ for $(2n+1)2\pi$ rotations could indeed clearly be verified (see Fig. 4.25b).

If both a nuclear phase change ϕ and a magnetic rotation α are introduced [4.128, 132, 133], characteristic interference and polarization effects arise. In this case $\hat{U} = \exp(i\phi + i\boldsymbol{\alpha\sigma}/2)$. Phase shift and polarization modulation can be worked out in a straightforward manner [4.132–134]. For unpolarized neutrons the intensity modulation of the 0-beam is

$$I_0(\phi, \alpha)/I(0, 0) = (1 + \cos\phi \cos\alpha/2)/2 . \tag{4.82}$$

The polarization of the 0-beam behind the interferometer is

$$\boldsymbol{P}_z(\phi, \alpha) = \alpha_u \sin\phi \sin\alpha/2(1 + \cos\phi \cos\alpha/2)^{-1} \tag{4.83}$$

with $\boldsymbol{\alpha}_u$ a unit vector parallel to the magnetic field $\boldsymbol{B}$. The experimental verification of the dependence of $I_0(\phi, \alpha)$ and $P_z(\phi, \alpha)$ by an interferometric experiment [4.134, 135] is shown in Fig. 4.25c. α_{eff} differs from α in that it takes into account the stray fields present also in the other interfering beam. Furthermore, in the experiment the intensity of the h-beam I_h was measured which is essentially in antiphase to I_0 modulated with a ϕ variation. As is seen, depending on the value of α_{eff}, either $I_h(\phi, \alpha)$ or $P_z(\phi, \alpha)$ shows maximum modulation with varying ϕ. Other interesting effects are expected if polarized incident neutrons are used [4.100].

In a very interesting experiment *Colella* et al. [4.136, 137] measured the phase shift suffered by the neutron wave function when its potential energy is altered along a beam path which is inclined against the horizontal. *Colella* et al. used an LLL neutron interferometer which they rotated about the incident beam by an angle Φ and measured the modulation of $(I_h - I_0)$. From a ray path theory they calculated the phase shift ϕ_g caused by gravity [4.138]:

$$\phi_g = 2\pi m_i m_g g h^{-2} \lambda A \sin \Phi . \tag{4.84}$$

Here m_i, m_g is the inertial and the gravitational mass of the neutron, respectively, g is the gravitational acceleration, and A is the area enclosed between the interfering beams. *Colella* et al. found a modulation consistent with ϕ_g within the experimental error, which, mostly due to simultaneous bending of the interferometer under rotation, was fairly large (10%). Earth's rotation causes an additional phase shift [4.138–140].

The fundamental importance of these experiments is related to the simultaneous occurence of both g and h in the above formula in a nontrivial combination.

4.6 Conclusions

Various types of Bragg diffraction interferometers have been proposed and some successfully operated. Among them the LLL-interferometer has reached the stage of being an X-ray and neutron optical instrument. Both polylithic and monolithic versions are applied.

Fringe formation, contrast, and imaging properties have been treated with a spherical wave theory including absorption which is capable of predicting numerically the intensity profiles of the outgoing beams.

Scattering factors for X-rays and coherent scattering lengths for neutrons are being measured with high precision. Furthermore, angstrom range interferometry is applied to phase contrast topography, crystal defect study, precision lattice parameter measurements, nuclear and magnetic scattering of neutrons by solid state structures, neutron spin rotation phenomena, and gravity-induced quantum interference.

Instrumental studies of other types of interferometers and for application to other fields such as phase determination in structure analysis, holography, and Fourier spectroscopy are going on. While the number of suggestions is increasing in these fields, no experimental results have been reached so far.

X-ray interferometry and neutron interferometry have definitely surpassed the instrumental stage, are contributing remarkable applications to other fields already and are expected to grow in importance steadily.

References

4.1   U.Bonse, M.Hart: Appl. Phys. Letters **6**, 155 (1965)
4.2   W.Bauspiess, U.Bonse, H.Rauch, W.Treimer: Z. Physik **271**, 177 (1974)
4.3   S.Tanemura, N.Kato: Acta Cryst. A**28**, 69 (1972)
4.4   L.L.Foldy: Phys. Rev. **87**, 693 (1952)
4.5   J.Schwinger: Phys. Rev. **73**, 407 (1948)
4.6   H.Maier-Leibnitz, T.Springer: Z. Physik **167**, 386 (1962)
4.7   U.Bonse: Present State of X-Ray Interferometry, in *Proceedings of the 5th Int. Congr. on X-Ray Optics and Microanalysis*, ed. by G.Moellenstedt, K.H.Gaukler (Springer Berlin, Heidelberg, New York 1969) pp. 1–10
4.8   U.Bonse: Neutron and X-Ray Interferometry. Int. Summer School on X-Ray Dynamical Theory and Topography, Limoges, ed. by F.Balibar (Laboratoire de Minéralogie et Cristallographie, Paris 1975)
4.9   M.Hart: Rep. Prog. Phys. **34**, 435 (1971)
4.10 M.Hart: Proc. R. Soc. A**346**, 1 (1975)
4.11 L.Marton: Phys. Rev. **85**, 1057 (1952)
4.12 L.Marton, J.Arol Simpson, J.A.Suddeth: Rev. Sci. Instr. **25**, 1099 (1954)
4.13 J.Arol Simpson: Rev. Sci. Instr. **25**, 1105 (1954)
4.14 G.Moellenstedt, H.Düker: Naturwiss. **42**, 41 (1954)
4.15 H.Düker: Z. Naturforsch. **10a**, 256 (1955)
4.16 G.Moellenstedt, H.Düker: Z. Physik **145**, 377 (1956)
4.17 H.Lichte, G.Moellenstedt, H.Wahl: Z. Physik **249**, 456 (1972)
4.18 M.Keller: Z. Physik **164**, 274 (1961)
4.19 H.Hoffmann, C.Jönsson: Z. Physik **182**, 360 (1965)
4.20 E.Kerschbaumer: Z. Physik **201**, 200 (1967)
4.21 B.Lischke: Phys. Rev. Letters **22**, 1366 (1969)
4.22 H.Boersch, B.Lischke: Z. Physik **237**, 449 (1970)
4.23 B.Lischke: Z. Physik **237**, 469 (1970)
4.24 B.Lischke: Z. Physik **239**, 360 (1970)
4.25 F.J.Landkammer: Z. Physik **189**, 113 (1966)
4.26 U.Bonse, M.Hart: Appl. Phys. Letters **7**, 99 (1965)
4.27 U.Bonse, M.Hart: Z. Physik **194**, 1 (1966)
4.28 U.Bonse, M.Hart: Acta Cryst. A**24**, 240 (1968)
4.29 M.Hart: Sci. Prog. Oxf. **56**, 429 (1968)
4.30 C.C.Chiu, H.Y.Chiu: Bull. Am. Phys. Soc. II.**21**, 667 (1976)
4.31 C.C.Chiu, H.Y.Chiu: Bull. Am. Phys. Soc. II.**21**, 668 (1976)
4.32 W.L.Bond, M.A.Duguay, P.M.Rentzepis: Appl. Phys. Letters **10**, 216 (1967)
4.33 R.M.J.Cotteril: Appl. Phys. Letters **12**, 403 (1968)
4.34 R.D.Deslattes: Appl. Phys. Letters **12**, 133 (1968)
4.35 M.A.Navasardyan: Izv. Akad. Nauk Arm. SSR Fiz. Mat. **8**, 187 (1973)
4.36 A.V.Kolpakov, R.N.Kuz'min, V.M.Ryaboy: J. Appl. Phys. **41**, 3549 (1970)

4.37   J.W.Giles: J. Opt. Soc. Am. **59**, 1179 (1969)
4.38   S.Aoki, Y.Ichihara, S.Kikuta: Jap. J. Appl. Phys. **11**, 1857 (1972)
4.39   S.Aoki, S.Kikuta: Jap. J. Appl. Phys. **13**, 1385 (1974)
4.40   N.Kato: Acta Cryst. **14**, 526 (1961)
4.41   N.Kato: Acta Cryst. **14**, 627 (1961)
4.42   D.Petrascheck: Acta Phys. Austriaca **45**, 217 (1976)
4.43   U.Bonse, M.Hart: Z. Physik **188**, 154 (1965)
4.44   D.Petrascheck, R.Folk: phys. stat. sol. (a) **36**, 147 (1976)
4.45   W.Bauspiess, U.Bonse, W.Graeff: J. Appl. Cryst. **9**, 68 (1976)
4.46   M.Kuriyama: Acta Cryst. A**27**, 273 (1971)
4.47   M. von Laue: *Röntgenstrahlinterferenzen*, 3rd ed. (Akademische Verlagsgesellschaft Frankfurt a. M. 1960)
4.48   E.Fermi: Ric. Sci. **7**, 13 (1936)
4.49   N.Kato: J. Appl. Phys. **39**, 2225 (1968)
4.50   N.Kato: J. Appl. Phys. **39**, 2231 (1968)
4.51   M.Ashkin, M.Kuriyama: J. Phys. Soc. Japan **21**, 1549 (1966)
4.52   N.Kato: Spherical Wave Theory. Int. Summer School on X-Ray Dynamical Theory and Topography, Limoges, ed. by F.Balibar (Laboratoire des Minéralogie et Cristallographie, Paris 1975)
4.53   H.Wagner: Z. Physik **146**, 127 (1956)
4.54   T.Saka, T.Katagawa, N.Kato: Acta Cryst. A**29**, 192 (1973)
4.55   B.W.Battermann, G.Hildebrandt: phys. stat. sol. **23**, K 147 (1967)
4.56   W.Graeff, U.Bonse: Z. Physik B**27**, 19 (1977)
4.57   H.Rauch, W.Treimer, U.Bonse: Phys. Letters **47A**, 369 (1974)
4.58   H.Rauch: Acta Phys. Austriaca **33**, 50 (1971)
4.59   U.Bonse, M.Hart: Z. Physik **190**, 455 (1965)
4.60   U.Bonse, E. te Kaat: Z. Physik **243**, 14 (1971)
4.61   U.Bonse, E. te Kaat: Z. Physik **214**, 16 (1968)
4.62   P.Becker, U.Bonse: J. Appl. Cryst. **7**, 593 (1974)
4.63   W.Bauspiess, U.Bonse, W.Graeff: Acta Cryst. A**31**, S 253 (1975)
4.64   H.Rauch, M.Suda: phys. stat. sol. A**25**, 495 (1974)
4.65   S.Kikuta, I.Ishikawa, K.Kohra, S.Hishino: J. Phys. Soc. Japan **39**, 471 (1975)
4.66   A.Authier: C. R. Acad. Sci. Paris **251**, 2003 (1960)
4.67   V.L.Indenbom, I.Sh.Slobodetskii, K.G.Truni: Sov. Phys. JETP **39**, 542 (1975)
4.68   E.V.Suvorov, V.I.Polovinkina: JETP Lett. **20**, 145 (1975)
4.69   P.A.Besirganyan, F.O.Eiramdshyan, K.G.Truni: phys. stat. sol. A**20**, 611 (1973)
4.70   U.Bonse, W.Graeff: Acta Cryst. A**31**, S 254 (1975)
4.71   P.Spieker: Dissertation (University of Dortmund 1977)
4.72   P.Penning, D.Polder: Philips Res. Rep. **16**, 419 (1961)
4.73   U.Bonse, W.Graeff, G.Materlik: Rev. Physique Appl. **11**, 83 (1976)
4.74   M.Hart: Phil. Mag. **26**, 821 (1972)
4.75   G.Christiansen, L.Gerward, A.Lindegaard Andersen: J. Appl. Cryst. **4**, 370 (1971)
4.76   L.Gerward, G.Christiansen, A.Lindegaard Andersen: Phys. Letters **39**, 63 (1972)
4.77   L.Gerward: Z. Physik **259**, 313 (1973)
4.78   U.Bonse, M.Hart, G.H.Schwuttke: phys. stat. sol. **33**, 361 (1969)
4.79   F.O.Eiramdshyan, P.A.Besirganyan: Izv. Akad. Nauk Arm. SSR Fiz. Mat. **7**, 215 (1972)
4.80   A.R.Lang, V.F.Miuscov: Appl. Phys. Letters **7**, 214 (1965)
4.81   J.Bradler, A.R.Lang: Acta Cryst. A**24**, 246 (1968)
4.82   G.H.Schwuttke, K.Brack: Z. Naturforsch. **28a**, 654 (1973)
4.83   C.Hoffrogge, H.J.Rademacher: PTB-Mitteilungen **83**, 79 (1973)
4.84   R.D.Deslattes: Bull. Am. Phys. Soc. **14**, 632 (1969)
4.85   R.D.Deslattes: Appl. Phys. Letters **15**, 386 (1969)
4.86   R.D.Deslattes: "Optical Interferometry of the 220 Repeat Distance in a Silicon Crystal" in *Precision Measurement and Fundamental Constants*, ed. by D.N.Langenberg, B.N.Taylor (NBS Special Publication 343, Gaithersburg 1971) pp. 279–283

4.87 M.Hart: Brit. J. Appl. Phys. 1, 1405 (1968)
4.88 U.Bonse, E.te Kaat, P.Spieker: "Precision Lattice Parameter Measurement by X-Ray Interferometry" in *Precision Measurement and Fundamental Constants*, ed. by D.N.Langenberg, B.N.Taylor (NBS Special Publication 343, Gaithersburg 1971) pp. 291–295
4.89 R.D.Deslattes, A.Henins: Phys. Rev. Letters 31, 972 (1973)
4.90 I.Curtis, I.Morgan, M.Hart, A.D.Milne: "A New Determination of Avogadro's Number" in *Precision Measurement and Fundamental Constants*, ed. by D.N.Langenberg, B.N.Taylor (NBS Special Publication 343, Gaithersburg 1971) pp. 285–289
4.91 M.Hart, J.G.Morgan: "Lattice Spacing Measurements and Avogadro's Number" in *Atomic Masses and Fundamental Constants 4*, ed. by J.H.Sanders, A.H.Wapstra (Plenum Press, London, New York 1972) Part 12, pp. 509–515
4.92 R.D.Deslattes, A.Henins, H.A.Bowman, R.M.Schoonover, C.L.Carroll, I.L.Barnes, L.A.Machlan, L.J.Moore, W.R.Shields: Phys. Rev. Letters 33, 463 (1974)
4.93 R.D.Deslattes, A.Henins, R.M.Schoonover, C.L.Carroll, H.A.Bowman: Phys. Rev. Letters 36, 898 (1976)
4.94 R.D.Deslattes: to be published in *Metrology and Fundamental Constants*, Vol. 68 (1977)
4.95 G.Basile: Alta Freq. 44, 549 (1975)
4.96 T.Fukamachi, S.Hosoya: Acta Cryst. A31, 215 (1975)
4.97 J.J.Sukarai: *Advanced Quantum Mechanics*, (Addison Wesley Reading, Mass. 1973)
4.98 U.Bonse, G.Materlik: Z. Physik B24, 189 (1976)
4.99 J.M.Cowley: *Diffraction Physics* (North Holland Publishing Company, Amsterdam, Oxford 1975) pp. 129–130
4.100 H.Rauch, G.Badurek, W.Bauspiess, U.Bonse, A.Zeilinger: "Determination of Scattering Lengths and Magnetic Spin Rotations by Neutron Interferometry" in *Proceedings of the Int. Conf. on the Interactions of Neutrons with Nuclei*, ed. by E.Sheldon, Vol. II (University of Lowell, Lowell, Mass. 1976) pp. 1027–1041
4.101 W.Marshall, S.W.Lovesey: *Theory of Thermal Neutron Scattering* (Clarendon Press, Oxford 1971) Chaps. 1, 10
4.102 C.Stassis, C.G.Shull: Phys. Rev. B5, 1040 (1972)
4.103 L.L.Foldy: Rev. Mod. Phys. 30, 471 (1958)
4.104 R.M.Thaler: Phys. Rev. 114, 827 (1959)
4.105 U.Bonse, H.Hellkötter: Z. Physik 223, 345 (1969)
4.106 U.Bonse, G.Materlik: Z. Physik 253, 232 (1972)
4.107 U.Bonse, G.Materlik: "Dispersion Correction Measurements by X-Ray Interferometry" in *Anomalous Scattering*, ed. by S.Ramaseshan, S.C.Abrahams (Munksgaard, Copenhagen 1975) pp. 107–109
4.108 D.C.Creagh, M.Hart: phys. stat. sol. 37, 753 (1970)
4.109 C.Cusatis, M.Hart: "Dispersion Correction Measurements by X-Ray Interferometry" in *Anomalous Scattering*, ed. by S.Ramaseshan, S.C.Abrahams (Munksgaard, Copenhagen 1975) pp. 57–68
4.110 C.Cusatis, M.Hart: Proc. Roy. Soc. A354, 291 (1977)
4.111 U.Bonse, G.Materlik, W.Schröder: J. Appl. Cryst. 9, 223 (1976)
4.112 W.Bauspiess, U.Bonse, H.Rauch: "The Perfect Crystal Neutron Interferometer: A Tool for Novel and Precise Measurement" in *Proceedings of the Conf. on Neutron Scattering*, ed. by R.M.Moon (National Technical Information Service, Springfield, Virginia 1976) pp. 1094–1102
4.113 W.Bauspiess, U.Bonse, W.Graeff, H.Rauch: J. Appl. Cryst. 10, 338 (1977)
4.114 F.O.Eiramdshyan, T.O.Eiramdshyan, P.A.Besirganyan: Izv. Akad. Nauk Arm. SSR Fiz. Mat. 9, 477 (1974)
4.115 D.C.Creagh: Australian J. Phys. 28, 543 (1975)
4.116 A.O.Aboyan, F.O.Eiramdshyan, P.A.Besirganyan: Izv. Akad. Nauk Arm. SSR Fiz. Mat. 9, 193 (1974)
4.117 N.Kato, S.Tanemura: Phys. Rev. Letters 19, 22 (1967)
4.118 J.Bohlen: Dissertation (University of Dortmund 1976)

4.119 M. Ando, S. Hosoya: "An Attempt at X-Ray Phase Contrast Microscopy" in *Proceedings of the Sixth Int. Conf. on X-Ray Optics and Microanalysis*, ed. by G. Shinoda, K. Kohra, T. Ichinokawa (University of Tokyo Press 1972) pp. 63–68

4.120 C. G. Shull, K. W. Billman, F. A. Wedgwood: Phys. Rev. **153**, 1415 (1967)

4.121 W. B. Dress, P. D. Miller, J. M. Pendlebury, P. Perrin, N. F. Ramsey: Phys. Rev. D **15**, 9 (1977)

4.122 H. Rauch, A. Zeilinger, G. Badurek, A. Wilfing, W. Bauspiess, U. Bonse: Phys. Letters **54A**, 425 (1975)

4.123 S. A. Werner, R. Colella, A. W. Overhauser, C. F. Eagen: Phys. Rev. Letters **35**, 1053 (1975)

4.124 Y. Aharonov, L. Susskind: Phys. Rev. **158**, 1237 (1967)

4.125 H. J. Bernstein: Phys. Rev. Letters **18**, 1102 (1967)

4.126 G. C. Hegerfeldt, K. Kraus: Phys. Rev. **170**, 1185 (1968)

4.127 G. T. Moore: Am. J. Phys. **38**, 1177 (1970)

4.128 A. G. Klein, G. I. Opat: Phys. Rev. D **11**, 523 (1975)

4.129 E. Drope: Z. Naturforsch. **29a**, 1117 (1974)

4.130 B. de Facio, P. W. Dennis, Y. P. Sharma: (to be published)

4.131 E. Klempt: Phys. Rev. D **13**, 3125 (1976)

4.132 A. Zeilinger: Z. Physik B **25**, 97 (1976)

4.133 G. Eder, A. Zeilinger: Nuovo Cimento **34B**, 76 (1976)

4.134 G. Badurek, H. Rauch, A. Zeilinger, W. Bauspiess, U. Bonse: Phys. Rev. D **14**, 1177 (1976)

4.135 G. Badurek, H. Rauch, A. Zeilinger, W. Bauspiess, U. Bonse: Phys. Letters **56A**, 244 (1976)

4.136 A. W. Overhauser, R. Colella: Phys. Rev. Letters **33**, 1237 (1974)

4.137 R. Colella, A. W. Overhauser, S. A. Werner: Phys. Rev. Letters **34**, 1472 (1975)

4.138 S. A. Werner, R. Colella, A. W. Overhauser, C. F. Eagen: "Neutron Interferometry" in *Proceedings of the Conf. on Neutron Scattering*, ed. by R. M. Moon (National Technical Information Service, Springfield, Virginia 1976) pp. 1060–1073

4.139 L. A. Page: Phys. Rev. Letters **35**, 543 (1975)

4.140 J. Anandan: submitted to Phys. Rev. (1976)

5. Section Topography

A. Authier

With 36 Figures

The use of nearly perfect crystals is generalizing rapidly in modern technology. A thorough characterization of their structural defects is absolutely necessary because of the interaction of these defects with most of the properties of these crystals. Dislocations, stacking faults, twin boundaries, subgrain boundaries, growth bands and striations, microdefects, long range strains, etc., are very important and all of them can be characterized non-distinctively by X-ray or neutron topography. All topographic methods are based on the difference between the reflecting powers of perfect and imperfect regions and on the propagation properties of X-rays or neutrons in slightly deformed areas. It is not therefore the defects themselves which are observed, but the strain fields they induce in the crystal. Both the position of the defects and the components of their strain-tensors can thus be determined. This is usually done by taking several topographs with different reflections. For instance, it is possible to find out the Burgers vector of a dislocation.

Although the first paper dealing with a topographic technique dates back to 1931 [5.1], this method of investigation really started around 1958 when the three main topographic methods (reflection, transmission, double crystal) were used for the first time to observe individual dislocations. Since then the topographic techniques have been increasingly used and more than 500 publications were devoted to studies performed on more than 80 different crystal types. There is actually quite a variety of topographic methods with different and complementary fields of applications (for reviews, see [5.2–7]) and the information content of X-ray topographs is multiplied by the use of combinations of topographic techniques and careful interpretation of images. The purpose of this chapter is to discuss one of these methods, *section topography*. It must be stressed that if topographic methods are in general simple in their principles, contrast problems are complicated and require a good knowledge of the dynamical theory of X-ray diffraction for their interpretation. Some of the bases of the dynamical theory of diffraction by perfect and nearly perfect crystals will therefore be given in this chapter.

The possibilities of topography are now being tremendously increased by the advent of neutron topography, of new powerful sources (see Chap. 2 of this volume) of synchrotron radiation and of new detectors enabling live topography of dynamic events such as dislocation movements or phase transformations (see Chap. 6 of this volume).

5.1 Principles of X-Ray and Neutron Topography

5.1.1 Basic Topographic Methods

The basic topographic methods are *Berg-Barrett topography* [5.8–10] and *projection topography* [5.11, 12]. The former is a reflection technique and the latter a transmission one. They both provide a general view of the distribution of defects; the main origins for the contrast are *misorientation contrast* (two regions are misoriented with respect to one another), *extinction contrast* (an imperfect region reflecting X-rays according to the geometric theory imbedded in a perfect region reflecting X-rays according to the dynamical theory), and *phase shifts* (such as those associated with stacking faults). The widths of dislocation images are roughly inversely proportional to the linewidth for the perfect crystal. They can be as narrow as a few microns or less for strong reflections, long wave lengths, and well-chosen asymmetric reflections.

Crystal thicknesses which can be studied with transmission topography vary from a few hundred microns for absorbing materials to several millimeters or more for lightly absorbing crystals such as silicon or organic crystals.

Crystal depths contributing to the image formation in reflection topography depend on absorption, the asymmetry of the reflection, the structure factor, the wavelength, etc. They can be varied at will between a fraction of a micron and 10 or 20 µm, which can be very useful, in particular in the study of epitaxial layers.

The possibility of obtaining a stereo effect by taking two pictures with hkl and $\overline{hkl}$ reflections or with two orientations of the crystal around the diffraction vector is extremely useful both in transmission [5.12–14] and in reflection topography [5.15].

5.1.2 Section Topography

Section topography is actually the oldest topographic method by which individual dislocations were resolved [5.16]. Its principle is recalled in Fig. 5.1. The incident beam coming from a point focus ($\sim 100 \times 100 \, \mu m^2$-optical) is collimated by a very narrow slit F_1 (width 5 to 10 µm). In a perfect crystal, it generates a fan of wave fields propagating within the so-called *Borrman* triangle ABC limited by the incident and reflected directions AB and AC, respectively. These wave fields give rise to interference phenomena (Pendellösung fringes) which have been interpreted by *Kato* using the spherical-wave dynamical diffraction theory [5.17–19]. They will be described in Section 5.3. If a defect such as a dislocation line D is present in the crystal, it gives rise to several images. One, the *direct image i_1* corresponds simply to a reflection of the direct beam on the distorted area around the defect [5.3]. It can easily be seen that the depth of the defect in the crystal can be deduced from the distance between the image i_1 and the edges of the diffraction pattern. One obtains thus an image of the defects contained in a *section* of the crystal by the sheet of incident X-rays, hence the name of *section topograph* or *section pattern* given by *Lang* [5.16]. When the crystal and the photographic plate are traversed simultaneously along direction

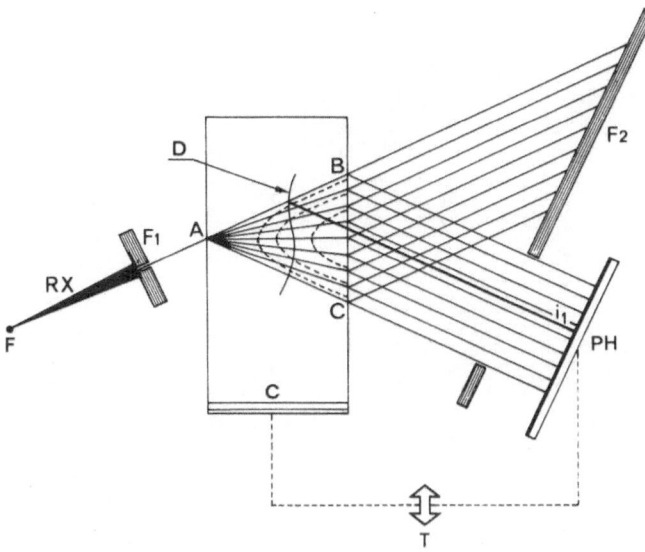

Fig. 5.1. Principle of section topography. F: focus; F_1 and F_2: first and second slits; PH: photographic plate; D: dislocation; i_1: direct image; T: translation of the crystal and photographic plate to obtain a traverse topograph

T, an image is obtained of the defects contained in the whole crystal volume. These images will lie on a projection of the defects along the reflected direction. This is one of the methods used to obtain a projection topograph, the *Lang method* [5.11, 12]. The pattern obtained is called a *traverse pattern* or *traverse topograph*. Of course, the quantitative information about the depth of the defects within the crystal is lost during the translation, and only qualitative information can be obtained, by looking at a stereo pair for instance.

Section topography is a complementary tool to projection topography. In practice, it will be performed on a spectrograph designed for traverse topography. Traverse topographs using different reflections and stereo pairs will first be taken. Section topographs will then be taken at positions chosen in advance on the corresponding traverse pattern. In general, the spectrographs designed for *Lang*'s method have a device giving the position of the traversing stage, and it is not very difficult to localize accurately the position of the section pattern relative to the traverse pattern. In some cases it is interesting to superimpose on the photographic plate a traverse topograph and a section topograph. A very good correspondence between the images of the defects on both topographs can thus be obtained. Figure 5.2 gives an example.

The presence of the defects also gives rise to perturbations of the propagation of wave fields within the *Borrmann* fan ABC. The effects related to the propagation of wave fields in deformed crystals will be described in Sections 5.4 and 5.5. They have been studied extensively, both theoretically and experimentally (for a review, see [5.3, 20–22]), but from a fundamental point of view. It has

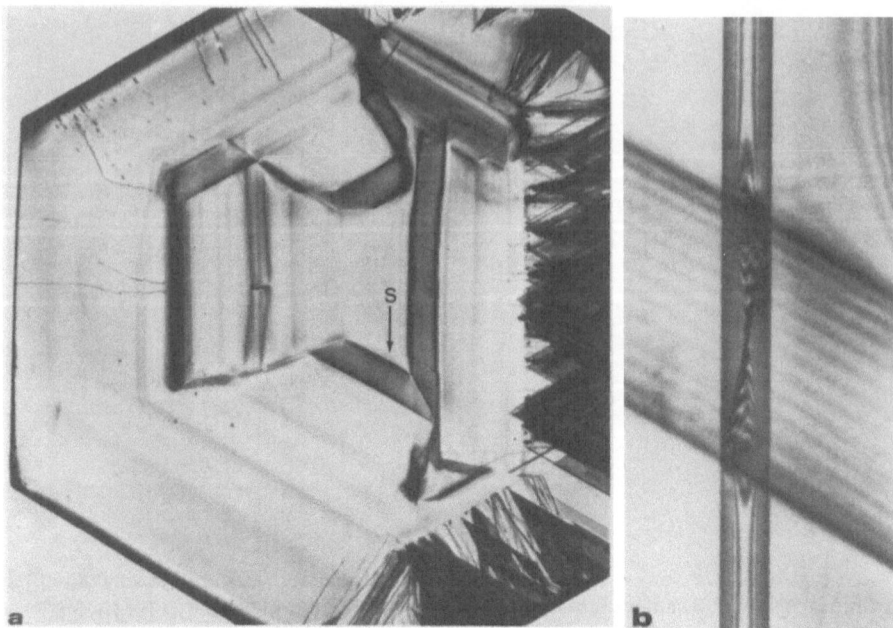

Fig. 5.2a and b. X-ray topographs of a quartz crystal (courtesy of *Zarka*, private communication). (a) Traverse topograph — Mo K_α. X 10; *S*: position of the section topograph of (b) superposition of a section and a traverse topograph. The defect observed is a growth band

not always been realized by users of topography for characterization that section topography is actually a very powerful tool for the assessment of crystal perfection. Let us give two examples besides the simple determination of the position of a defect.

1) The visibility of Pendellösung fringes on a section pattern depends critically on the perfection of the crystal and decreases strongly in the presence of microdefects or microinclusions distributed throughout the crystal, which would not give rise to any image on a projection topograph, giving the wrong impression of crystal perfection. This effect has been used for instance by *Patel* to follow the precipitation of oxygen during annealing of Czochralski grown silicon crystals [5.23] (see Fig. 5.3).

2) A planar defect creates a surface of discontinuity between two parts of the crystal and therefore a disruption in the propagation of wave fields and a characteristic diffraction pattern which is different on a section pattern for the following different cases:

— stacking fault [5.24, 25]
— twin boundary between regions having different structure factors [5.20, 26]
— misorientation boundary [5.20, 26, 27]
— layer of microinclusions generating a strain gradient on each side of the fault [5.28]

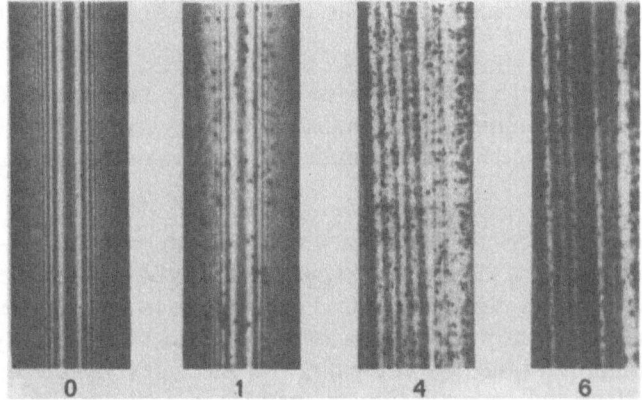

Fig. 5.3. X-ray section topographs of a silicon crystal (after [5.23]). The crystal contains
$7.5 \times 10^{17}\, \mathrm{cm^{-3}}$ atoms of oxygen. As the crystal is annealed at 1000 °C, the oxygen precipitates, and
the visibility of Pendellösung fringes decreases. The numbers given under each topograph
correspond to the number of annealing hours

It will be shown in Section 5.6 how it is possible to distinguish these different
cases on a section pattern. It is much more difficult to do so on a traverse pattern
on which sets of parallel fringes are observed in each case.

5.1.3 Double Crystal Topography

Double crystal topography is essentially used
    I) in the $(+, -)$ setting, in the *Bragg-Bragg* arrangement [5.29–31], or in the
Bragg-Laue arrangement [5.32–34];
    II) in the $(+, +)$ setting, in the *Bragg-Laue* arrangement [5.34].
    In the $(+, -)$ or nondispersive setting, the diffraction vectors of the two
successive reflections are parallel and of opposite senses. In the $(+, +)$ or
dispersive setting, the diffraction vectors of the two successive reflections are not
exactly parallel but have the same sense.
    The $(+, -)$ setting provides a very high sensitivity to small strains and has
been used quite extensively for this reason. Let us mention, for instance, highly
accurate local lattice parameter measurements [5.35].

5.1.4 Special Techniques

Many variations of the techniques mentioned above have been developed for
special purposes and in particular to enhance the contrast of defects such as
microdefects whose contribution to the general contrast is usually small. Let us
mention briefly, among others
    I) *limited projection topography* [5.36, 37] in which a traverse topograph is
taken after occulting part of the reflected beam with one or both lips of slit F_2

(Fig. 5.1); the direct images of defects lying in part of the crystal volume can thus be eliminated;

II) *direct beam topography* [5.36] in which a traverse topograph is taken using the refracted beam after eliminating the reflected beam and the edge *AB* of the direct beam. All direct images are thus eliminated and dynamical contrast enhanced;

III) *kinematical image topography* [5.38] in which a double crystal $(+, -)$ *Laue-Laue* arrangement is used. The two crystals are adjusted so that their reflecting planes are exactly parallel. A slit placed between the two crystals lets only the *direct beam AB* go through. This beam is deprived of all the rays satisfying *Bragg*'s condition for the perfect regions of the second crystal. When traversing this second crystal only the direct images are recorded on the photographic plate; their contrast is thus enhanced;

IV) *topographs set on the flank of the rocking curve.* This technique is of course the standard one in $(+, -)$ *Bragg-Bragg* crystal topography and it enables one to determine the sign and magnitude of small strains in a situation of dynamical contrast [5.30, 39]. It can also be used in ordinary transmission topography to study the sign of the misorientations associated with direct images [5.40]. If a topograph is recorded while the crystal is set at a small percentage of the peak intensity ($\sim 5\%$) in the $(+, -)$ *Bragg-Laue* double crystal arrangement, for instance, the dynamically diffracted intensity is considerably reduced and the contrast of direct images enhanced. This has been applied to the study of microdefects such as swirls [5.41].

5.1.5 Neutron Topography

One of the drawbacks of X-ray topography is that samples should not be too thick because of absorption. In comparison, the advantage of neutron topography is that for most crystals absorption is very much smaller than in the case of X-rays. Much larger crystals and, in particular, as-grown crystals, can thus be investigated. There are two main problems with neutron topography. One is related to the relatively low neutron intensities, even for high-flux reactors. The other one is resolution. There are three limitations to resolution in neutron topography.

I) *Instrumental Limitations.* Neutron beams are broad. It is therefore necessary to collimate them very severely. When neutron guides can be used, losses are greatly reduced.

II) *Photographic Limitations.* There is no direct photographic detector and a conversion to photons or to α-particles is necessary. In every case there is a resolution loss.

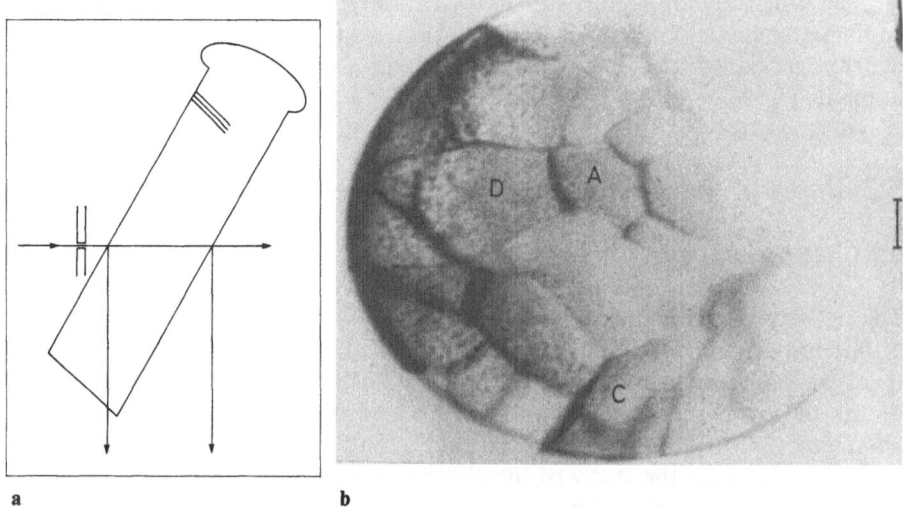

a   b

Fig. 5.4a and b. Neutron section topograph of an uncut as-grown Si 3 % Fe crystal (after [5.44]). (a) Experimental setting. (b) Neutron section topograph. Diameter of the crystal : 20 mm $\lambda = 2.8$ Å ; 011 ; scale mark : 2 mm

III) *Theoretical Limitations.* As has been mentioned earlier, the width of the direct image of a dislocation is inversely proportional to the width of the rocking curve for the perfect crystal. This width is smaller for neutrons than for X-rays, and dislocation images are therefore broader on neutron topographs.

For all these reasons, the narrowest dislocation images in neutron topography are of the order of 30 µm. It is possible nevertheless to obtain clearly resolved images of individual dislocations [5.42].

Neutron beams are polychromatic. There are therefore two possible types of experimental settings. One consists simply of working with the white beam and recording one or several of the resulting *Laue* spots. The other one, as in most neutron diffraction experiments, consists of using a monochromated incident beam. This technique, which may be less convenient than the former one in the case of crystals made of subgrains, has less background and a better resolution.

As in the case of X-rays, both projection and section topographs can be recorded, the former with a wide entrance slit after the monochromator (parallel beam topography) and the latter with a narrow one (Fig. 5.4a). Because of the nature of the neutron spectrum, the problem of the $K_{\alpha_1} - K_{\alpha_2}$ doublet for X-rays does not exist here. The first application of neutron section topography is due to *Tomimitsu* and *Doi* [5.43] who were studying the strain field in a severely hot pressed germanium crystal, but the real possibilities of the technique were demonstrated by *Schlenker* and co-workers [5.44] (Fig. 5.4b). The distribution of defects in a *section* of a large crystal can thus be investigated without having to cut the crystal.

The other advantage of neutron topography is of course the possibility of studying magnetic domains as has been shown for instance, in the case of spin density waves [5.45–47] and ferromagnetic [5.48] and antiferromagnetic domains [5.49].

5.2 Bases of the Dynamical Theory of X-Ray Diffraction

5.2.1 Introduction

The bases of the dynamical theory were laid very soon [5.50, 51] after the first diffraction experiment of X-rays by a crystal [5.52]. At first only the plane wave theory was developed and it was not applied to a large extent. The development of the theory really started in the 1940s and 1950s with the discovery of anomalous absorption, or *Borrmann* effect [5.53–55] and its interpretation [5.56, 57] and with the study of the X-ray energy propagation [5.58–63]. The discovery of techniques for growing perfect single crystals such as silicon and germanium and the development of the topographic techniques induced further studies of the propagation of real waves inside the crystal. This led to the evaluation of the energy distribution on a section pattern [5.64, 65]. Actually, real waves are not plane, and the classical dynamical theory had to be generalized to account for the propagation of spherical waves [5.18, 19] or any kind of waves [5.66]. The most important concept in dynamical theory is that of *wave fields* and it will be shown how it can explain the absorption and propagation properties of X-rays in a perfect crystal and the contrast of defect images. For a detailed discussion of the dynamical theory, the reader may refer to [5.20, 67, 68].

5.2.2 The Propagation Equation

There are two possible starting points to derive the dynamical theory. One is to use the *Fresnel* zone calculation and *Darwin*'s recurrence formulae [5.50]; this was first done for the *Bragg* case but has been extended to the *Laue* case [5.69]. The other, and more widely used, starting point is given by *Maxwell*'s equations. The foundations of the corresponding theory were laid by *Ewald* [5.51], who considered a crystalline medium as a discrete three-dimensional lattice of dipoles. The theory was further developed by *Laue* [5.67, 70] who considered the crystalline medium as a *continuous* distribution of negative charges. It can easily be shown that the corresponding dielectric susceptibility is equal to

$$\chi = -\frac{e^2}{mc^2}\frac{1}{4\pi\varepsilon_0}\frac{\lambda^2}{\pi}\varrho = -R\lambda^2\varrho/\pi \tag{5.1}$$

where ϱ is the electronic density, λ the wavelength, and R is the classical radius of the electron.

In a crystal, the electron density ϱ is triply periodic. It can therefore be expanded in *Fourier* series and the dielectric susceptibility as well.

$$\chi = \sum_h \chi_h \exp(2\pi i \, h \cdot r) \quad \text{with} \quad \chi_h = -R\lambda^2 F_h/\pi V. \tag{5.2}$$

h is a reciprocal lattice vector, F_h the corresponding structure factor, V the volume of the unit cell, and $\sum_h$ a triple summation. The *Fourier* coefficients χ_h are of the order of 10^{-6} and negative.

If the crystal is absorbing, the dielectric susceptibility is complex

$$\chi = \chi^r + i\chi^i. \tag{5.3}$$

The imaginary part χ^i is also triply periodic and can be expanded in *Fourier* series

$$\chi = \sum_h^i \exp(2ih \cdot r)$$

with

$$\chi_0^i = -\mu/2\pi k \tag{5.4}$$

(μ is the linear absorption coefficient). The *Fourier* coefficients of the imaginary part χ_h^i are in general small relative to those of the real part χ_h^r.

The electric field, the electric displacement, the magnetic field, and the magnetic induction are related at each point by *Maxwell*'s equations

$$\text{curl}\, \mathbf{E} = -\frac{\partial \mathbf{B}}{\partial t} \quad \text{curl}\, \mathbf{H} = \frac{\partial \mathbf{D}}{\partial t}$$

$$\text{div}\, \mathbf{D} = 0 \qquad \text{div}\, \mathbf{B} = 0. \tag{5.5}$$

The local charge and current density are taken to be equal to zero. The justification for the assumed local neutrality is that since the nuclei do not contribute to the scattering, their charge may be expanded over the volume occupied by the electron cloud to compensate the negative charge.

$$E_{T_1} - E_{T_2} = 0 \quad D_{N_1} - D_{N_2} = 0$$

$$H_{T_1} - H_{T_2} = 0 \quad B_{N_1} - B_{N_2} = 0. \tag{5.6}$$

T and N indicate tangential and normal components, respectively.

The electric field and displacement, the magnetic field and induction are related by

$$\mathbf{D} = \varepsilon \mathbf{E} = \varepsilon_0 (1 + \chi)\mathbf{E}; \quad \mathbf{B} = \mu_0 \mathbf{H}; \quad \varepsilon_0 \mu_0 = 1/c^2. \tag{5.7}$$

ε_0 is the dielectric permittivity and μ_0 the magnetic permittivity of vacuum.

We shall throughout consider incident monochromatic waves with frequency v,

$$E \exp(2\pi i v t). \tag{5.8}$$

The propagation equation is deduced from (5.5) and (5.7) by eliminating E, H, and B.

$$\Delta D + \operatorname{curl} \operatorname{curl} \chi D + 4\pi^2 k^2 D = 0 \tag{5.9}$$

with $k = v/c$.

5.2.3 Plane Wave Theory

In a medium for which χ is constant, (5.9) reduces to

$$\Delta D = 4\pi^2 k^2 (1 + \chi) D \tag{5.10}$$

of which the simplest solution is a plane wave

$$D = D_0 \exp[2\pi i(v t - K \cdot r)] \tag{5.11}$$

with $K = k \sqrt{1 + \chi} = nk$ (n index of refraction). If χ is very small, $K = k(1 + \chi/2)$.
The general solution of (5.9) is a linear combination of partial solutions, a combination which is determined by the boundary conditions. It can be shown that the simplest particular solution for a plane wave incident on a semi-infinite crystal is given, when χ is triply periodic, by a *Bloch wave* having the same triple periodicity:

$$D_j = \exp[2\pi i(v t - K_0 \cdot r)] \sum_h \mathcal{D}_{hj} \exp(2\pi i h \cdot r). \tag{5.12}$$

Subscript j denotes the particular solution. Equation (5.12) can also be written

$$D_j = \exp(2\pi i v t) \sum_h \mathcal{D}_{hj} \exp(-2\pi i K_{hj} \cdot r) \tag{5.13}$$

with

$$K_{hj} = K_{0j} - h. \tag{5.14}$$

This *Bloch* wave is interpreted as a wave field consisting of an infinite number of plane waves with wave vectors K_{hj} and amplitudes $\mathcal{D}_{hj}$. In practice only a small number of them have a nonnegligible amplitude and we shall consider only the usual situation where this number is equal to two. All the wave vectors of a given wave field, drawn from the various reciprocal lattice points, define a *tie-point* P_j, characteristic of the wave field (Fig. 5.5).

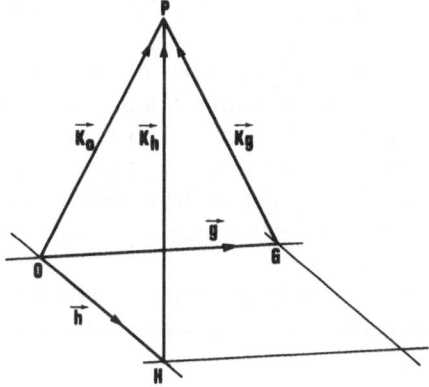

Fig. 5.5. *P*: tie-point characterizing a given wave field. *O, H, G*: reciprocal lattice points

The notion of wave field is a very important one because all their wave components have common properties which permit a complete description of the energy propagation in a perfect or nearly perfect crystal. For example, in an absorbing crystal, the wave vectors are complex. Equation (5.14) shows that the wave vectors of a given wave field all have the same imaginary part since the reciprocal lattice vectors *h* are real

$$K^i_{hj} = K^i_{0j}. \tag{5.15}$$

All the waves belonging to the same wave field thus undergo the same absorption. They also have the same propagation direction. It is given by the Poynting vector. Its time average over a period is equal to

$$\bar{S}_j = \tfrac{1}{2} \mathscr{R}[E_j \times H^*_j] \tag{5.16}$$

where $\mathscr{R}(z)$ means real part of z.

The expressions of E_j and H^*_j can be deduced from (5.12) and *Maxwell's* equations (5.5). One finds when there are two terms only in expansion (5.12)

$$\bar{S}_j = \frac{c}{\varepsilon_0} \exp(4\pi K^i_{0j} \cdot r)(s_0 \mathscr{D}_{0j} + s_h \mathscr{D}_{hj}) \tag{5.17}$$

where s_0 and s_h are unit vectors in the direction of K_0 and K_h, respectively.

Let us now substitute the expansions (5.2) and (5.12) into the propagation equation (5.9). It now consists of an infinite number of terms and it is possible to show that it is equivalent to an infinite system of linear homogenous equations

$$\mathscr{D}_{hj} - \frac{K^2_{hj}}{K^2_{hj} - k^2} \sum_{h'} \chi_{h-h'} \mathscr{D}_{h'[h]j} = 0 \tag{5.18}$$

where the summation h' is over all reciprocal lattice points, $\chi_{h-h'}$ is the *Fourier* coefficient associated with reciprocal lattice vector $h - h'$, and $\mathscr{D}_{h'[h]j}$ means the component of $\mathscr{D}_{h'j}$ on the plane normal to K_h. Equations (5.18) were called the fundamental equations of dynamical theory by v. *Laue*.

The unknowns of the system (5.18) are the amplitudes $\mathscr{D}_{hj}$ of the waves belonging to a given wave field. For the system to have nontrivial solution (that is, for the wave field to exist), its determinant should be equal to zero. This provides an equation which connects the values of the distances K_{hj} between the various reciprocal lattice points and the tie-point P_j. This is the equation of a surface in reciprocal space, called *the dispersion surface: a wave field can exist only if its tie-point lies on the dispersion surface.*

In the two-beam case, that is when there are two reciprocal lattice points only on the *Ewald* sphere 0 and H, that is when expansion (5.12) is reduced to two terms, the system (5.18) is reduced to

$$2X_{0j}\mathscr{D}_{0j} - k\chi_{\bar{h}}^- C\mathscr{D}_{hj} = 0$$
$$-k\chi_h C\mathscr{D}_{0j} + 2X_{hj}\mathscr{D}_{hj} = 0 \tag{5.19}$$

where $C=1$ when the polarization is normal to the plane K_0, K_h and $C=\cos 2\Theta$ when the polarization is parallel to this plane. X_{0j} and X_{hj} are the distances from the tie-point P_j to the two spheres centered in 0 and H, respectively, and with radii $k(1+\chi_0/2)$.

The ratio of the amplitudes of the two waves which constitute a wave field is equal to:

$$R_{hj} = \frac{\mathscr{D}_{hj}}{\mathscr{D}_{0j}} = \frac{k\chi_h C}{2X_{hj}} = \frac{2X_{0j}}{k\chi_{\bar{h}}C}. \tag{5.20}$$

The equation of the dispersion surface may now be written, dropping the subscript j,

$$X_0 X_h = \frac{k^2}{4}\chi_h\chi_{\bar{h}}C. \tag{5.21}$$

It is a surface of revolution around $\overline{OH}$ as axis and it is made of the two spheres of centers O and H and radii $k(1+\chi_0/2)$ and a connecting surface near the intersection of the two spheres. Only this connecting surface is interesting since it corresponds to the case where both amplitudes, $\mathscr{D}_{0j}$ and $\mathscr{D}_{hj}$, are nonnegligible. Figures 5.6 and 5.7 represent the intersection of the dispersion surface with a plane passing through O and H. Figure 5.7 is a close-up view. Since the *Fourier* coefficients χ_h and $\chi_{\bar{h}}$ are very small, the distances X_0 and X_h of P to the two spheres are very small compared to the radius of the spheres, and on Fig. 5.7 these spheres are replaced by their tangential planes. Equation (5.21) shows that the intersections of the dispersion surface with a plane passing through O and H is a hyperbola asymptotic to the tangents T_0 and T_h to the two spheres.

The propagation direction of both waves of a given wave field is given by the *Poynting* vector (5.17). Using (5.19) and (5.21), it can be shown that this

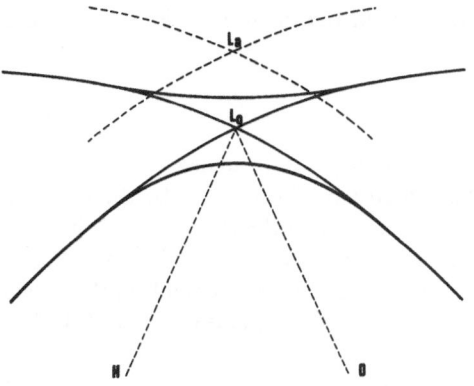

Fig. 5.6. Dispersion surface. La: *Laue* point. It lies at the intersection of the two spheres with radii $k = 1/\lambda$; Lo: *Lorentz* point. It lies at the intersection of the two spheres centered at O and H, respectively, with radii $k(1 - \chi_0/2)$

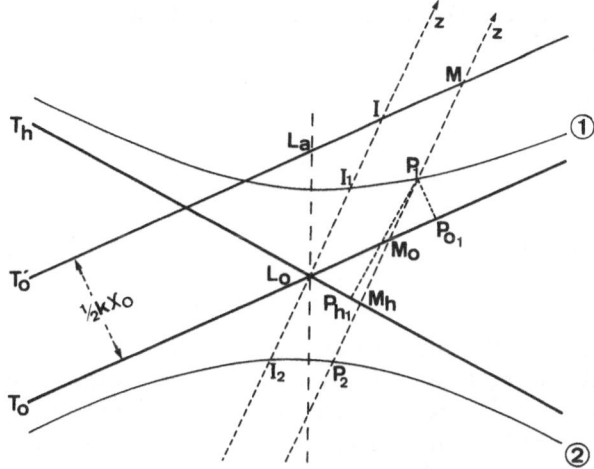

Fig. 5.7. Dispersion surface: close-up view—plane-wave case

propagation direction is *normal to the dispersion surface* at the tie-point characteristic of the wave field. The angle α_j between this direction and the lattice plane is given by

$$\text{tg}\,\alpha_j = \frac{1 - R_{hj}^2}{1 + R_{hj}^2}\,\text{tg}\,\Theta$$

(5.22)

where Θ is the *Bragg* angle and R_{hj} is defined by (5.20).

The above discussion tells us which wave fields may propagate in a crystal and the *boundary conditions* for the wave-vectors tell us *which particular ones* are excited in a given problem. The boundary conditions for the amplitudes enable us to calculate the absolute values of the amplitudes of the waves in each wave field.

The boundary condition for the wave vectors is that of the *continuity of the tangential component* of the wave vectors at the crystal surfaces. Let $K_0^{(a)} = \overline{OM}$ be the wave vector of the plane wave incident on the crystal. M lies on a sphere centered at O and with radius

$$K_0^{(a)} = k = 1/\lambda. \tag{5.23}$$

In the vicinity of the interesting portion of the dispersion surface (Fig. 5.7), it can be replaced by its tangential plane T_0'. Let Mz be the normal to the entrance surface of the crystal: it cuts the dispersion surface at the tie-points of the wave fields excited in the crystal by the incident plane wave. Figure 5.7 shows that in the transmission case an incident plane wave excites two wave fields in the crystal for each direction of polarization. The tie-points of these two wave fields belong to both branches of the dispersion surface.

Let La or *Laue* point be the intersection of the two circles centered in O and H and with radii $k = 1/\lambda$ (Fig. 5.6). The plane wave of wave vector $\overline{O\,La}$ satisfies the geometric *Bragg* conditions exactly. We shall designate the departure from *Bragg*'s law by the angle (see Fig. 5.7)

$$\Delta\Theta = (\overline{O\,La}, \overline{OM}) = \overline{LaM}/k. \tag{5.24}$$

We shall further call

$$\begin{aligned}
\psi_0 &= (Mz, \overline{OM}) \ ; \quad \cos\psi_0 = \gamma_0 \\
\psi_h &= (Mz, \overline{HM}) \ ; \quad \cos\psi_h = \gamma_h.
\end{aligned} \tag{5.25}$$

Let Lo or *Lorentz* point be the intersection between the tangents T_0 and T_h to the circles centered in O and H, respectively, and with radii $k(1 + \chi/2)$ and let I, I_1, and I_2 be, respectively, the intersections of the parallel to Mz drawn from Lo with T_0', and the two branches of the dispersion surface. $\overline{OI}$ is the wave vector of the incident plane wave corresponding to the middle of the reflection domain (or of the rocking curve). We shall define the following quantities:

$$\Delta\Theta_0 = \overline{LaI}/k \tag{5.26}$$

$$\Lambda_0 = 1/I_1 I_2 = \sqrt{\gamma_0\gamma_h}/k|C|\sqrt{\chi_h\chi_{\bar{h}}} \tag{5.27}$$

("Pendellösung" fringe distance)

$$\begin{aligned}
\eta &= k\frac{\gamma_0}{\gamma_h}(\Delta\Theta - \Delta\Theta_0)\sin 2\Theta \\
&= \frac{\Delta\Theta \sin 2\Theta + \frac{1}{2}k\chi_0(1 - \gamma_h/\gamma_0)}{|C|\sqrt{\chi_h\chi_{\bar{h}}}\sqrt{\gamma_h\gamma_0}}
\end{aligned} \tag{5.28}$$

(deviation parameter).

The values of the amplitudes of the waves which constitute each wave field are obtained by combining (5.6) and (5.19). Since an incident plane wave excites two wave fields of tie-points P_1 and P_2, these wave fields interfere, generating an oscillating term in the expressions of the total reflected and refracted intensities. The phase difference between the two wave fields is equal to

$$2\pi(\overline{OP}_1 - \overline{OP}_2) \cdot r = 2\pi \overline{P_1 P_2} \cdot r . \tag{5.29}$$

If we notice that, by construction, $\overline{P_1 P_2}$ is normal to the crystal surface, one finds for a plane parallel slab of thickness t

$$2\pi P_1 P_2 \cdot t = 2\pi \frac{t}{\Lambda} . \tag{5.30}$$

It can be shown that

$$\Lambda = \Lambda_0 / \sqrt{1 + \eta^2} . \tag{5.31}$$

This interference effect has been called *Pendellösung* by *Ewald*.

The rocking curve contains therefore oscillations which are quickly damped as absorption increases, as we shall see in Section 5.2.4. The envelope of the rocking curve is a curve representing a function proportional to $1/(1 + \eta^2)$. The half width of the rocking curve is therefore equal to

$$\delta \eta = 2$$

that is, using (5.28) and (5.2), to

$$\begin{aligned}
\delta &= 2|C| \sqrt{\chi_h \chi_{\bar{h}}} \sqrt{\gamma_h / \gamma_0} / \sin 2\Theta \\
&= 2|C| R \lambda^2 F_h \sqrt{\gamma_h / \gamma_0} / \pi V \sin 2\Theta .
\end{aligned} \tag{5.32}$$

This expression is of great practical importance in topography. It can be seen that the width of the rocking curve is larger for a longer wavelength, a stronger structure factor, and asymmetric reflections for which $\psi_h < \psi_0$.

5.2.4 Anomalous Transmission

As we mentioned earlier, absorption is taken into account phenomenologically by introducing an imaginary part to the dielectric susceptibility. This has as a consequence that the wave vectors and the dispersion surface are complex. The absorption factor in the expression of the intensity of a wave comes in through a term which can be written

$$\exp(2\pi K_0^i \cdot r) = \exp(-\mu_a t / \gamma_0) \tag{5.33}$$

where μ_a is the apparent absorption coefficient and t the crystal thickness. If one computes the value of μ_a using the properties of the dispersion surface, one obtains, in the case of a symmetric reflection ($\gamma_0 = \gamma_h = \cos\Theta$),

$$\mu_a = \mu(1 \mp |C| \, |\chi_h^i/\chi_0^i| \cos\phi/\sqrt{1+\eta^2}) \tag{5.34}$$

where μ is the normal absorption coefficient and ϕ the phase difference between the *Fourier* coefficients χ_h^r and χ_h^i. The top sign corresponds to branch 1 of the dispersion surface and the bottom one to branch 2.

Equation (5.34) shows that the effective absorption coefficient is different from the normal one in the reflection domain (η different from $\pm\infty$), that it is smaller for branch 1 and larger for branch 2 (when $\cos\phi > 0$). This effect is maximum when $\eta = 0$; it is called *anomalous transmission* and was discovered by *Borrmann* [5.53, 54]. It is of great practical importance: because of it the X-ray intensity can propagate at the *Bragg* angle through much larger crystal thicknesses than otherwise. Furthermore, Pendellösung interference effects between wave fields belonging to both branches of the dispersion surface are quickly damped when the crystal thickness increases. The intensity of the *Borrmann* effect depends on the ratio $|\chi_h^i/\chi_0^i|$; it can be very close to 1 in simple structures (for instance, for reflections with even indices for silicon or germanium) but this is not the case for complicated structures.

Borrmann has given a simple physical interpretation of the anomalous transmission effect. He considers the interference between the two waves constituting a given wave field. The intensity of the wave field can be written, in the two beam case, from [5.12].

$$|D_j|^2 = |\mathscr{D}_{0j}|^2 [1 + |R_{hj}|^2 + 2R_{hj}C\cos(2\pi h \cdot r)] . \tag{5.35}$$

It can be seen that a set of stationary waves is formed. The nodes lie on planes of equation

$$h \cdot r = \text{const} \tag{5.36}$$

that is parallel to the lattice planes. They lie *on* the lattice planes for $R_{hj} < 0$, that is for wave fields belonging to branch 1, while the *antinodes* corresponding to wave fields belonging to branch 2 also lie on the lattice planes. There is a minimum of electric field at the atoms for branch 1 wave fields and therefore very little absorption, but the situation is reversed for branch 2.

5.2.5 Spherical Wave Theory

In practice, real waves incident on a crystal are never plane waves. There are two main ways to study the diffraction of waves different from a plane wave.

I) *Generalization of the Fundamental Equations of the Dynamical Theory.* For any kind of wave other than a plane wave or if the crystal is not semi-infinite, or if there are strains present in the crystal, it is no more possible to consider a solution of the propagation equation (5.9) having the simple form (5.13) where the amplitudes and the wave vectors are constant. If one assumes that the amplitudes are slowly varying functions, it is possible to keep the expression (5.13) with constant wave vectors but with varying amplitudes. This has been done by *Takagi* [5.66, 71] and *Taupin* [5.72]. The system of linear equations (5.18) is then replaced by a system of partial differential equations,

$$\frac{\partial \mathscr{D}_0}{\partial s_0} = - i\pi k C \chi_{\bar{h}} \mathscr{D}_h$$

$$\frac{\partial \mathscr{D}_h}{\partial s_h} = - i\pi k (C \chi_h \mathscr{D}_0 - 2\beta_h \mathscr{D}_h)$$

(5.37)

where

$$\beta_h = \frac{|K_h|^2 - |K_0|^2}{2k^2} \sim \frac{K_h - k(1 + \chi_0/2)}{k}.$$

(5.38)

The constant wave vectors in expansion (5.13) are chosen at will and in the most convenient manner relative to the boundary conditions at the entrance surface. Generally K_0 is taken to be equal to $k(1 + \chi_0/2)$ and $K_h = K_0 - h$.

The parameter β_h is constant in the case of a perfect crystal and it is quite possible, using appropriate boundary conditions, to choose K_h in such a way that $\beta_h = 0$. In a deformed crystal, however, β_h varies from point to point and is related to the components of the strain tensor by the expression [5.71]

$$\beta_h = \frac{1}{k} \frac{\partial}{\partial s_h} (h \cdot u)$$

(5.39)

where u is the displacement vector associated with the deformation.

This expression is directly related to be local variation $\delta\Theta$ of the departure from *Bragg*'s law (or *effective misorientation*) due to the deformation [5.73]:

$$\delta\Theta = \frac{-1}{\sin\Theta} \beta_h = - \frac{1}{k \sin 2\Theta} \frac{\partial(h \cdot u)}{\partial s_h}.$$

(5.40)

In simple cases, for instance a perfect crystal and an incident spherical wave, system (5.37) can be solved analytically [5.71] and the intensity distribution along the base of the *Borrmann* triangle is proportional to

reflected beam $\quad R_h : I_h = A_0 J_0'^2 \left(\pi \frac{t}{\Lambda_0} \sqrt{1 - Y^2} \right),$

(5.41)

refracted beam $\quad R_0 : I_0 = A_1 \frac{1 + Y}{1 - Y} J_1^2 \left(\pi \frac{t}{\Lambda_0} \sqrt{1 - Y^2} \right)$

(5.42)

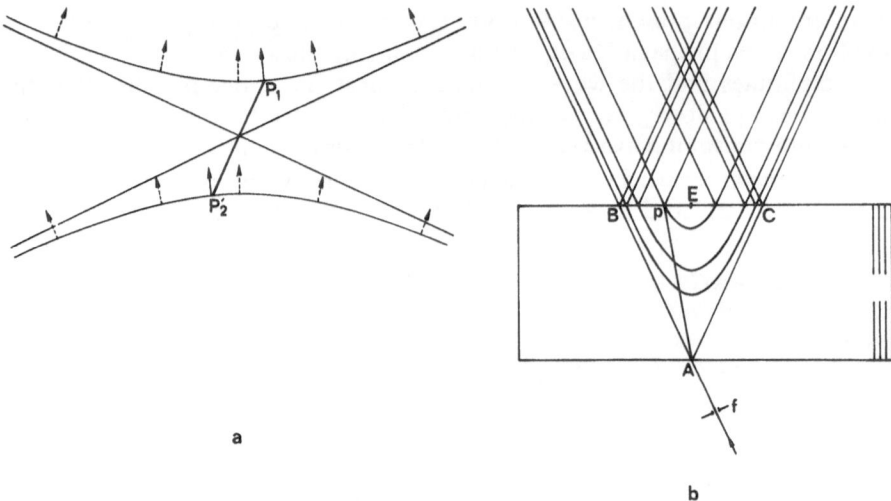

a

b

Fig. 5.8a and b. Spherical wave case, (a) dispersion surface, (b) direct space

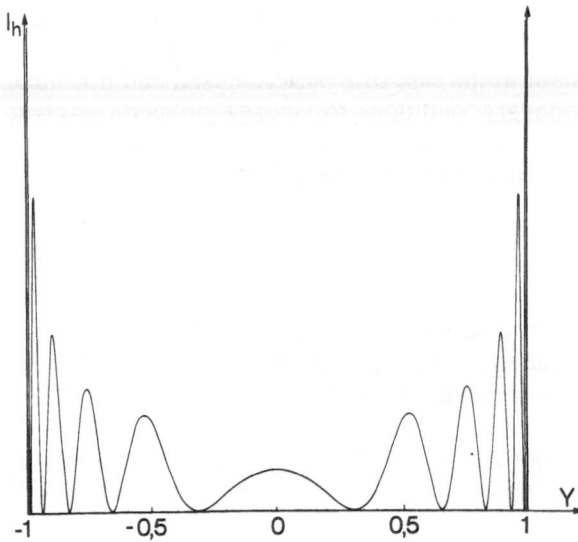

Fig. 5.9. Intensity distribution along the base of the *Borrmann* triangle for the reflected beam in the spherical wave case (*Kato*-Pendellösung fringes)

where Λ_0 is given by (5.27) and $Y = \overline{Ep}/\overline{EC}$, E being the midpoint of $\overline{BC}$ (Fig. 5.8) and p a point along BC. Figure 5.9 represents the intensity distribution for the reflected beam and a particular value of t/Λ_0.

In more complicated cases, in particular in the case of a deformed crystal, system (5.37) can be solved numerically only with computer aid.

II) *Expansion of the Incident Wave by a Fourier Transform.* Expressions (5.41) and (5.42) were first obtained by *Kato* [5.18] and extended by him to the

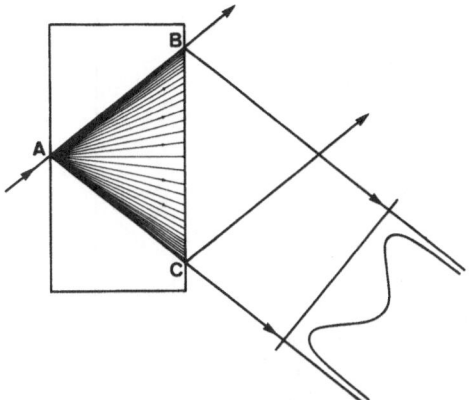

Fig. 5.10. Fanning out of the wave fields and intensity distribution in the reflected beam, neglecting Pendellösung, in the medium μt range

absorbing case [5.19]. He showed that an incident spherical wave could be expanded, by means of a *Fourier* transform, in a distribution of plane wave components having all the same wave number. To obtain the intensity distribution, he applied the plane wave dynamical theory to each plane wave component, then calculated its *Fourier* transform. This *Fourier* transform can be reduced to a one-dimensional transform in the plane of incidence where the reciprocal variables are, on one hand, the deviation parameter η of the plane wave components and, on the other hand, the reduced coordinate Y of a point along the base of the *Borrmann* triangle in the spherical wave.

Parseval's theorem shows that the intensity integrated along the base of the *Borrmann* triangle on a section pattern, that is the intensity received by a point of the photographic plate in a traverse pattern, is equal to the integrated intensity under the rocking curve in the plane wave theory.

5.3 Discussion of the Intensity Distribution Across a Section Pattern

5.3.1 Intensity Distribution for Each Branch of the Dispersion Surface

When the incident wave can be considered to be a spherical wave, the angular width of its *Fourier* expansion in plane waves is always much larger than the width of the rocking curve in the plane wave case. This means that every point of the dispersion surface is excited and that the corresponding wave fields are generated and propagate inside the crystal. Their propagating directions spread out in the so-called *Borrmann* fan (Fig. 5.10) which was first described by *Borrmann* et al. [5.74]. *Kato* has calculated the corresponding intensity distribution $I(y)$ along the base of the *Borrmann* fan for each branch of the dispersion surface independently [5.65] (Fig. 5.11). It can be written

$$I_{m_j}(y) = I_0(\Delta\Theta)\frac{d(\Delta\Theta)}{dy}\frac{\gamma_m}{\gamma_0}|\mathscr{D}_m^{(d)}/\mathscr{D}_0^{(a)}|^2 \quad j=1,2; \quad m=0,h \qquad (5.43)$$

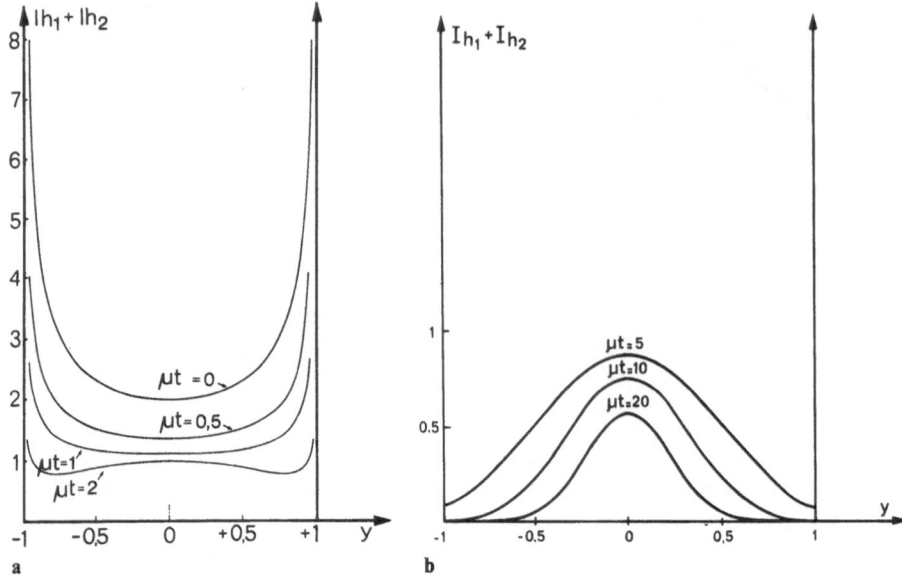

Fig. 5.11a and b. Intensity distribution along the base of the *Borrmann* triangle for the reflected beam, neglecting Pendellösung. (a) $\mu t = 0, 0.5, 1, 2$; (b) $\mu t = 5, 10, 20$

where $I_0(\Delta\Theta)$ is the angular distribution of the energy of the incident beam, and $\mathscr{D}_0^{(a)}$ and $\mathscr{D}_m^{(d)}$ are the amplitudes of the incident and reflected or refracted waves, respectively.

The ratio $\mathscr{D}_h^{(d)}/\mathscr{D}_0^{(a)}$ is given by the plane wave theory for each point of the dispersion surface. It is maximum at the center of the reflection domain. The ratio $d(\Delta\Theta)/dy$ is proportional to the radius of curvature of the dispersion surface. It is therefore very small at the middle of the reflection domain and very large far from the reflection domain where the dispersion surface is asymptotic to the two spheres of radius $k(1 + \chi_0/2)$ (Fig. 5.7). The result is that in the small absorption case, expression (5.43) is small in the middle of the base BC of the *Borrmann* triangle and increases asymptotically towards the edges B and C of the section pattern. This has been called *margin effect* by *Lang* and is one of the distinctive features of section patterns (see Fig. 5.10).

When μt increases, the influence of anomalous transmission becomes important: branch 2 wave fields are more and more absorbed and only branch 1 wave fields with their tie-point close to the apex of the hyperbola are little absorbed. Because of this effect a maximum appears in the middle of the intensity distribution along the base of the *Borrmann* triangle (Fig. 5.11b).

5.3.2 Pendellösung or Kato Fringes

If one considers a particular path Ap in the *Borrmann* fan (Fig. 5.8), it is readily seen that there are two tie-points of the dispersion surface corresponding to wave

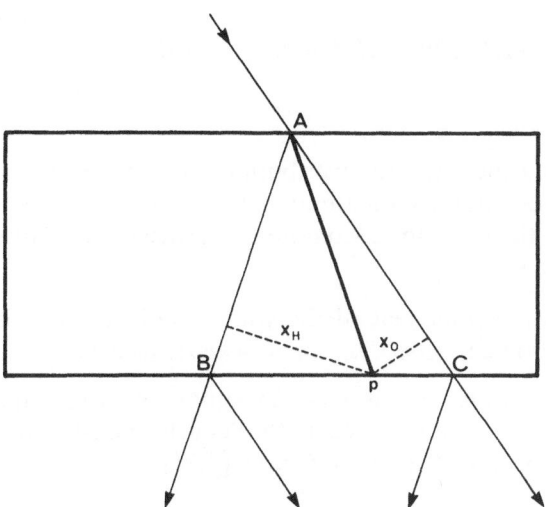

Fig. 5.12. Distances of a point on the exit surface from the sides of the *Borrmann* triangle

fields propagating along this path: P_1 and P'_2 which lie on a diameter of the dispersion surface. The two wave fields propagating along the same path Ap are coherent and interfere. The points of constant phase are such that

$$P_1 P'_2 \cdot Ap = \text{const.} \tag{5.44}$$

It can be shown by using the properties of the dispersion surface that

$$P_1 P'_2 \cdot Ap = t/\Lambda_0 \sqrt{1 + \eta^2} = t \sqrt{1 - Y^2}/\Lambda_0 \tag{5.45}$$

where η is the deviation parameter of P_1 and Y has been defined at the end of Section 5.2.5, p. 162. From this definition it is readily seen that $t^2(1 - Y^2)$ is proportional to the product $x_0 x_h$ of the distances of p from AB and AC (Fig. 5.12). The equal amplitude curves are therefore hyperbolae $x_0 x_h = \text{const}$ asymptotic to the sides of the *Borrmann* triangle (Fig. 5.8). This result was first observed by *Kato* and *Lang* [5.17] and obtained theoretically by *Kato* [5.18]. It is in agreement with (5.41) and (5.42) since the argument of the *Bessel* function is equal to $\pi t \sqrt{1 - Y^2}/\Lambda_0$ and the *Bessel* function J_0 can be approximated by

$$J_0(\xi) = \sqrt{\frac{2}{\pi \xi}} \cos\left(\xi - \frac{\pi}{4}\right). \tag{5.46}$$

The coefficient $(1 - Y^2)^{-1/2}$ which therefore appears in the intensity distribution (5.41, 42) corresponds to the geometrical effect described in the preceding subsection.

Parallel fringes are therefore expected in the section pattern of plane parallel plate when the value of μt is not too high and hyperbolic fringes when the crystal is wedge shaped. The visibility of these fringes decreases very rapidly when the crystal perfection decreases (Fig. 5.3) and their presence is therefore a good criterion of crystal perfection.

5.4 Propagation of Wave Fields in a Deformed Crystal

5.4.1 Introduction

The preceding section described the intensity distribution on a section topograph of a perfect crystal. This intensity distribution is modified when the crystal is strained. Although there is no discontinuity between them, one may distinguish two extreme cases.

I) *Small strain gradients* [the component of the strain gradient which is important is a quantity proportional to $\partial(\delta\Theta)/\partial s_0$ where $\delta\Theta$ is defined by (5.40)].

II) *Large strain gradients* which correspond to situations where the curvature of the lattice planes over a distance equal to the Pendellösung distance Λ_0 is much larger than the width δ of the rocking curve [5.75].

5.4.2 Small Strains

In the case of small strains the propagation of wave fields in the crystal may be calculated using theories based on geometric optics [5.76–78]. The basic assumption is that the perfect crystal theory applies locally and that the local deviation parameter of a given wave field varies according to the point to point variations of the effective misorientation $\delta\Theta$. One may thus assume that the tie-point characteristic of the wave field moves along the branch of the dispersion surface to which it belongs. The propagation direction and the phase of the tie-point vary accordingly. It can be shown in particular that when the strain gradient is constant the path of the wave fields is a hyperbola asymptotic to the sides of the *Borrmann* triangle (Fig. 5.13).

Penning and *Polder* [5.76] have discussed this situation in the high μt range where branch 2 wave fields are completely absorbed out. Associated with the curvature of the path of the wave fields there is a variation of the corresponding intensity in the reflected (R_h) and refracted (R_0) beams. *Penning* and *Polder* have shown that the curvature of the paths of the wave field belonging to branch 1 is in the same sense as that of the lattice planes, in the opposite direction for wave fields belonging to branch 2 (Fig. 5.14). There is a transfer of intensity between the two waves which constitute a given wave field.

Type 1 wave fields: From the refracted to the reflected wave when the center of curvature I lies on the same side of the lattice planes as point H of the reciprocal lattice (Fig. 5.14a). From the reflected to the refracted wave if the center of curvature and the reciprocal lattice point H lie on each side of the lattice planes (Fig. 5.14b).

Type 2 wave fields: Just the opposite conclusions apply.

Kato [5.77, 78] used a slightly different approach and, taking into account the phases, studied the interferences between wave fields. At each point p of the

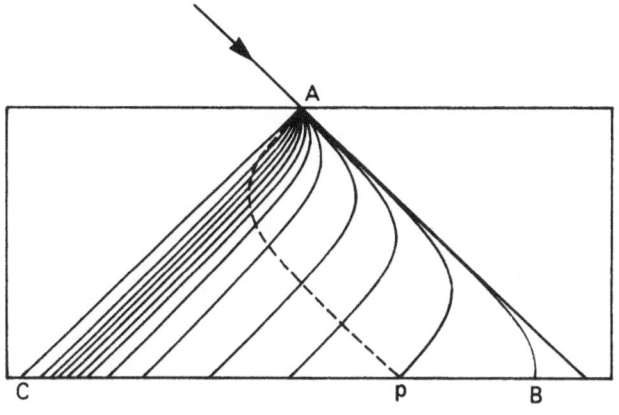

Fig. 5.13. Paths of the wave fields in the *Borrmann* triangle for various values of a constant strain gradient (full curves: branch 1, broken curve: branch 2) (after [5.20])

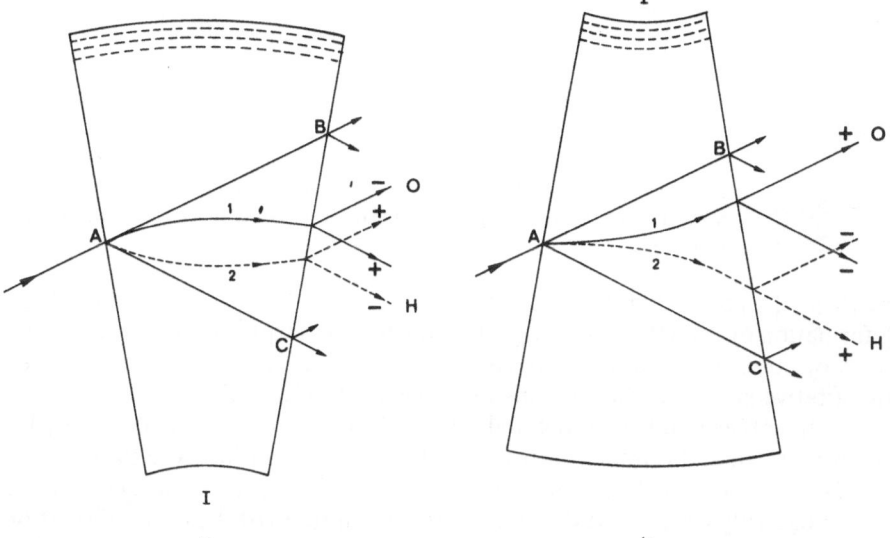

Fig. 5.14. Wave field paths and energy transfers in a curved crystal. 1, 2: type of wave field; $+$, $-$: increase or decrease of beam intensity

exit surface (Fig. 5.13) two wave fields arrive, as in the perfect crystal case. But they have now followed different paths having opposite curvatures. Two effects are thus expected.

On one hand, the phase difference between the two wave fields has increased because of the deformation and one expects a deformation of the Pendellösung fringes by contraction of the fringe spacing. The contraction increases as the strain gradient increases and new fringes appear (Fig. 5.15). This phenomenon was experimentally demonstrated by *Ando* and *Kato* [5.79] and more quanti-tatively by *Hart* [5.80] for a crystal having a temperature gradient. It has been applied to the quantitative study of the strain induced in a silicon crystal by the

Fig. 5.15. Section topograph of a triglycine sulphate crystal. Note the deformation of the Pendellösung fringes due to the presence of a strain gradient (after [5.117])

presence of an oxide film on part of its surface [5.81–83]. Let us note that if this deformation of Pendellösung fringes is at the origin of equal deformation fringes on projection topographs, section topographs are the best tools for the quantitative study of small strains as is shown by [5.80–83].

On the other hand, as in the high μt range, there are transfers of energy between reflected and refracted waves, but they are of opposite sense for branch 1 and branch 2 wave fields. The observed contrast is therefore complicated and cannot be easily predited. It depends on the magnitudes of the lattice distortion and of the anomalous transmission as has been discussed by *Ando* and *Kato* [5.84].

Another possibility to obtain the intensity distribution on a section pattern consists in solving (5.37) in the case of a deformed crystal. This can always be done by means of a computer [5.85] and sometimes analytically, for instance, in the case of a constant strain gradient [5.86]. However, no interpretation of the results in terms of ray trajectories can be made.

5.4.3 Large Strains

When the strain gradient is too large (see Sec. 5.4.1, p. 166), "ray optics" or "geometrical optics" are no more valid and diffraction effects of the propagating wave fields take place, exactly as in the case of a beam of light propagating in a region where the index of refraction varies very fast. This effect was first mentioned in *Penning*'s thesis [5.87] and discussed in detail by *Balibar* and co-workers [5.75, 88–91]. These diffraction effects can be described by viewing crystal deformations as resulting in secondary sources excited by each incoming wave field. These sources reemit wave fields belonging to *both* branches of the dispersion surface (*interbranch* scattering). Therefore, starting with one wave field, one is bound to obtain two packets of wave fields while going through a region of strong distortions [5.75]. This *creation* of new wave fields is the specific form that diffraction takes in the case of X-ray propagation; it occurs only when the deformations are large enough for diffraction effects to be observable. The criterion mentioned in Section 5.4.1, p. 166 is analogous to the well-known criterion for the validity of geometrical optics with ordinary light. This effect—creation of new wave fields in highly distorted areas—was used to interpret some characteristic features of dislocation images in section topographs [5.3] and has been demonstrated directly by studying the diffraction of an isolated wave field by the strain field surrounding a dislocation [5.92].

Takagi's [5.66, 71] and *Taupin*'s [5.72] theories which include the possibility of crystal deformations from the start are quite general and take into account diffraction effects. They imply the creation of new wave fields; this has been shown for instance in the case of large thermal gradients [5.91] or of dislocation images [5.93]. It is important to note that by solving (5.37) on a computer it is possible to obtain simulated dislocation images which agree very well with those observed on a section topograph [5.93]. The features of the image vary with the orientation and sign of the *Burgers* vector and it is possible to determine the *Burgers* vector by comparison of observed and simulated images.

We have mentioned in this section and shall describe in Section 5.5.4 the importance of interbranch scattering in the formation of dislocation images. It should be stressed that all highly strained regions will induce such interbranch scattering which will be made visible by enhancements of the intensity observed on the topographs. This can be the case for instance of surface damage: close to the damage there will be blackening due to new wave fields and further away there will be a deformation of the Pendellösung fringes. If there are micro-inclusions, their strain field may be large enough close to them to induce interbranch scattering, even if they are too small to be observed, and a general blackening will be observed.

5.5 Dislocation Images on a Section Topograph

5.5.1 Various Types of Images

Figure 5.16 shows the principle of the formation of dislocation images on section topographs. Let us consider a dislocation line intersecting the sides of the *Borrmann* triangle at D_1 and D_2. We have already noted in Section 5.1.2 that the reflection of the direct beam by the strained regions around D_1 gives rise to a *direct image* i_1.

Let us further consider the path AP of a particular wave field which passes very close to the dislocation at P. This path is going to be perturbated by two effects.

I) Close to the core of the dislocation, strain gradients are sufficiently large for interbranch scattering to take place: wave fields propagating along AP will excite wave fields propagating in a parallel direction along PM and *new* wave fields having a tie-point on the other branch of the dispersion surface and propagating along PQ. The latter will give rise to an image i_3, called *intermediary* image. Because of the energy which has been deviated in direction PQ, there will be a decrease of the intensity of the wave fields propagating along PM.

II) The paths of wave fields propagating through the less strained regions further away from the core are curved and are therefore deviated from their original direction [5.94].

The result of both these effects is a depletion in the direction APM of a wave field path intersecting a dislocation. This gives rise to a light shadow at i_2 on the photographic plate which is called the *dynamical* image. It is usually lined on each side by a black fringe [5.95–97].

The relative importance of the width and contrast of the three types of images depends on the position of the dislocation relative to the *Borrmann* triangle and its orientation, the crystal thickness, and the absorption.

5.5.2 Direct Images

The construction of Fig. 5.16 shows that the intensity of the direct image is superimposed on the intensity distribution corresponding to the perfect crystal and its contrast is black. We shall first discuss the depth of the dislocation and then its contrast.

I) *Depth of the Dislocation.* If the direct image of a defect lies on the edge b of the section topograph, it can be seen from Fig. 5.16 that the defect lies on the exit surface; if the defect lies on the edge c, the defect lies on the entrance surface. If the center of the image lies at a distance h from edge b, the defect is at a distance z from the exit surface,

$$z = h \cos \psi_0' / \sin 2\Theta \pm \Delta z \tag{5.47}$$

where the uncertainty Δz is related to the width l of the incident beam by

$$\Delta z = l \cos \psi_h' / \sin 2\Theta . \tag{5.48}$$

ψ_0' and ψ_h have been defined in Section 5.2.3. The width of the beam depends on the width f_1 of the entrance slit (Fig. 5.1) and on the divergence ω of the direct beam

$$l = f_1 + L\omega \tag{5.49}$$

where L is the distance of the slit from the crystal.

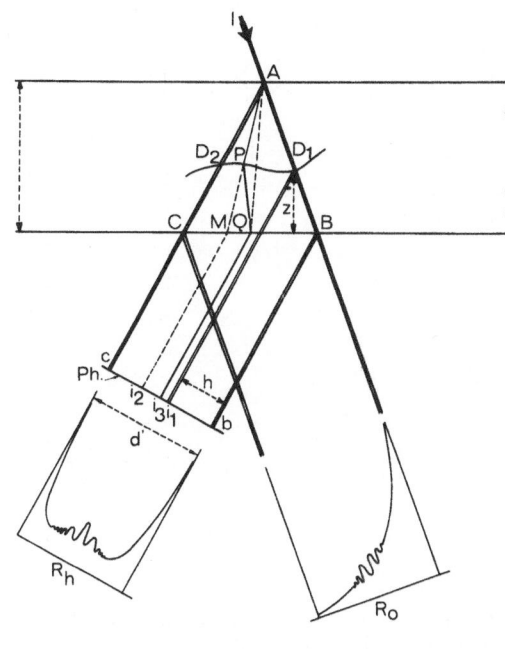

Fig. 5.16. Formation of dislocation images in section topographs. i_1: direct image; i_2: dynamical image; i_3: intermediary image

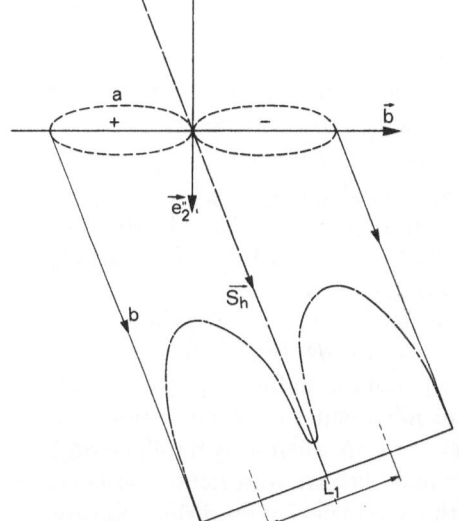

Fig. 5.17. Formation of direct images. a Equal effective misorientation curve around an edge dislocation. b: *Burgers* vector parallel to the surface; S_h: unit vector in the reflected direction; e_2'': unit vector normal to the surface. b Intensity distribution of the direct image

The accuracy on the determination of the position of the defect is therefore the better, the thinner the entrance slit and the smaller the divergence of the beam. For instance, if

$$f_1 = 5\,\mu\text{m}, \quad L = 4\,\text{cm}, \quad \omega = 2 \times 10^{-4},$$
$$\psi_h' = \Theta = 12°, \quad l = 13\,\mu\text{m}, \quad \Delta z \sim 30\,\mu\text{m}. \tag{5.50}$$

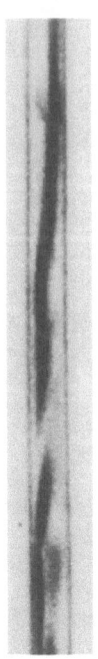

Fig. 5.18. Section topograph of a triglycine sulphate crystal. (200) reflection. Mo K_α. (courtesy of *Petroff*, private communication). Note the double contrast of direct images

The width of the image is in general much smaller, and the uncertainty on the position of the defect depends mainly on the geometric parameters.

II) *Contrast of Direct Images – Kinematical Model.* The principle of the kinematical model of direct dislocation images was given by *Lang* [5.13, 98] and *Wilkens* and *Meier* [5.99] and developed quantitatively by *Chikawa* [5.100] and *Authier* [5.3]. It has been applied numerically to mica [5.101], silicon [5.102], triglycine sulphate [5.3], and various other organic crystals [5.103].

The angular and spectral widths of the incident beam are always larger than the line width of the perfect crystal. The direct beam AD_1 (Fig. 5.16) is therefore constituted of rays which do not satisfy *Bragg*'s condition for the *perfect* crystal. In Fig. 5.17, *a* represents an equal effective misorientation curve around the dislocation corresponding to a value equal to $\pm x\delta$ where δ is the line width (5.32) and x a number close to one. This equal effective misorientation curve consists of two or more lobes depending on the parameters of the dislocation and of the reflection. The basic assumption of the model is that the regions situated inside these equal effective misorientation curves are sufficiently deformed relative to the perfect crystal to *Bragg* reflect the direct beam AD_1. The regions thus behave like a small ideally imperfect crystal imbedded in the perfect crystal and reflect X-rays according to the geometric or kinematical theory. The intensity distribution is therefore proportional to the volume of imperfect crystal. It is schematically represented by *b* in Fig. 5.17. Figure 5.18 gives an example.

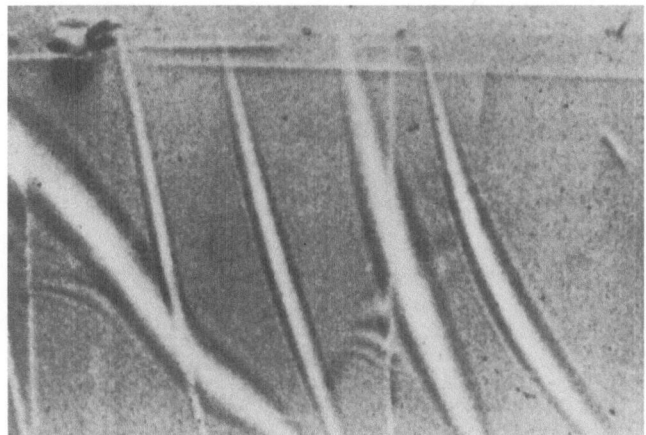

Fig. 5.19. Section topograph of a silicon crystal. Note the dynamical images of dislocations: white lines flanked by black fringes. (220) reflection $-$ Ag K_α, crystal thickness: 4 mm × 64

The model is of course only an approximation since there is no discontinuity between the *direct* and *intermediary* images. It is in practice difficult to separate the direct images from the other images and it is remarkable that the agreement with experimental results is quite good [5.3, 101–103]. The experimental values of x vary between 1 and 2 and depend on the relative importance of the three types of images, that is, on the parameters of the dislocation and the reflection and on the absorption.

5.5.3 Dynamical Images

I) *General Properties.* Figure 5.19 gives examples of dynamical images observed on a section topograph of a 4 mm thick silicon crystal. Several properties of the dynamical images have been shown experimentally [5.3, 16] and interpreted either by means of a model [5.96] or of computer simulations [5.93].

1) The contrast is identical in the reflected and refracted beams.

2) The contrast is identical on topographs taken with hkl and $\bar{h}\bar{k}\bar{l}$ reflections.

3) The width of dynamical images is the larger, the further the dislocation lies from the exit surface. The width becomes equal to the whole width of the *Borrmann* triangle when the dislocation reaches the entrance surface. It is therefore possible to estimate the depth of a dislocation from the width of its dynamical image. Examples can be seen on Fig. 5.19.

4) The width of the dynamical image of a dislocation lying at a constant depth is increased from the sides towards the center of the *Borrmann* triangle (that is towards the middle of the section topograph). This is quite clear in Fig. 5.19.

5) If a dislocation (or a scratch) lies on the exit surface, its dynamical image coincides with the projection of the defect on the plate. Its contrast is no more

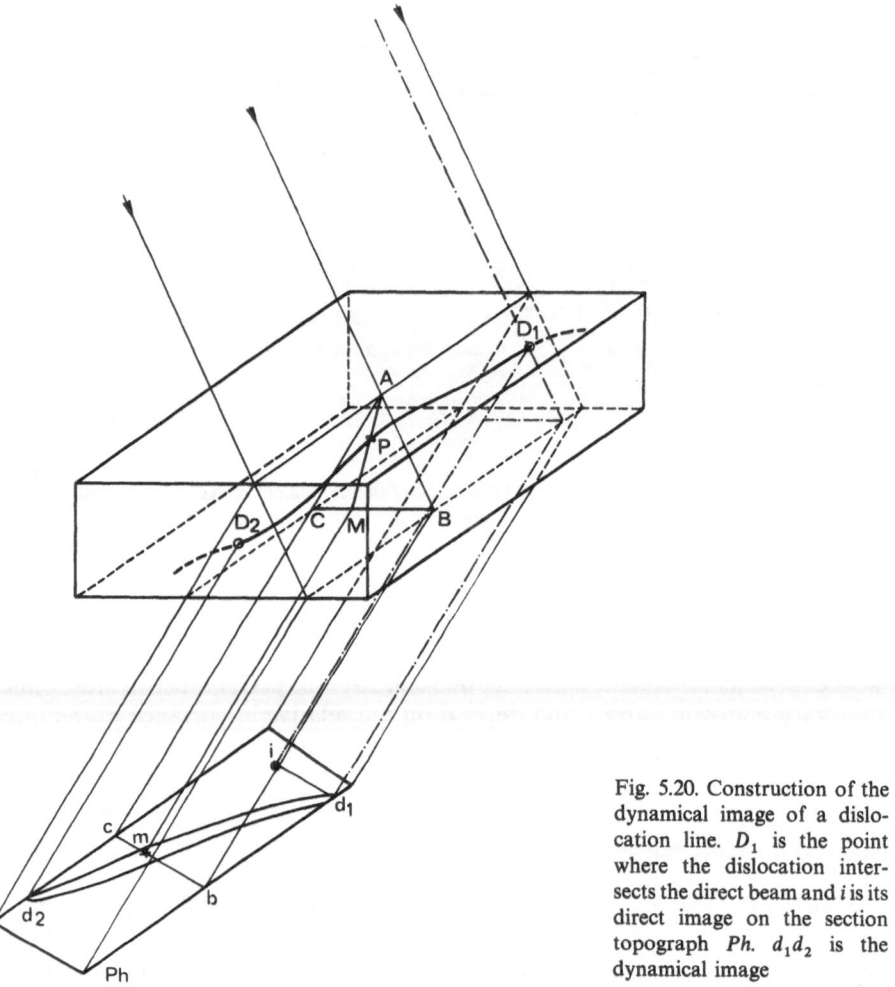

Fig. 5.20. Construction of the dynamical image of a dislocation line. D_1 is the point where the dislocation intersects the direct beam and i is its direct image on the section topograph Ph. d_1d_2 is the dynamical image

black-white-black, but double: black-white. It is reversed in the reflected and refracted beams and for hkl and $\bar{h}\bar{k}\bar{l}$ reflections.

II) *Geometrical Properties.* From what has been said in Section 5.5.1, the dynamical image can be considered as a shadow cast in the direction of the paths of the wave fields incident on the dislocation. Figure 5.20 shows how the dynamical and direct images of a dislocation line on a section topograph can be constructed.

The dynamical image of a straight dislocation image is curved if the dislocation is not parallel to the crystal surfaces. This is illustrated in Fig. 5.21. The sense of the curvature of the dynamical image changes when the sense of the slope of the dislocation also changes. The figure also shows that if one can superpose the direct image of a dislocation on a projection topograph and its

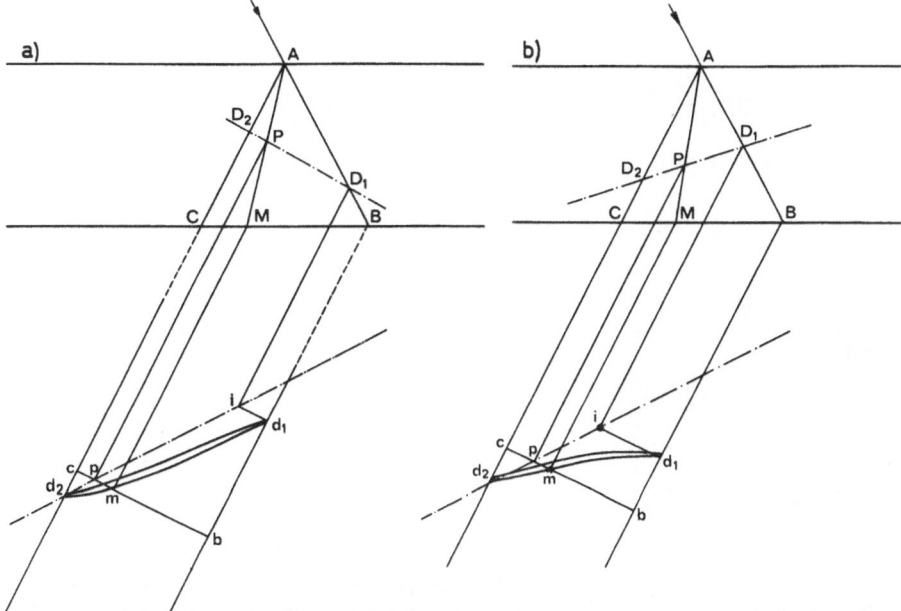

Fig. 5.21. Dependence of the shape of the dynamical image on the indication of the line. The top of each diagram is a frontal projection on the plane of incidence and the bottom part a projection on the photographic plate. The projections of the dislocation are drawn in broken lines. The projection on the photographic plate is the same for both figures but the frontal projections are different. Note that the curvature of the dynamical image is also different. bc: trace of the incidence plane ABC on the photographic plate; m: dynamical image of P; p: intersection of bc with the direct image of the dislocation on a projection topograph: it is the projection of P along the reflected direction

dynamical image on a section topograph, the exact position of the dislocation in the crystal can be reconstructed. Let P be a point of the dislocation line and ABC the plane of incidence passing through P. The dynamical image of P on the section topograph is m, projection on the photographic plate of the intersection M of AP with the base of the Borrmann triangle. The direct image of P on the projection topograph is simply its projection p in the reflected direction. Conversely P can be determined from p and m.

5.5.4 Intermediary Images

We have mentioned in Section 5.4.3 that when wave fields propagating in the crystal are incident on highly distorted regions they induce new wave fields whose tie-points lie on the other branch of the dispersion surface (interbranch scattering). This occurs for instance at the vicinity of a dislocation core. If AP is the path of the incident wave field, the new wave fields will propagate along a different direction, PQ (Fig. 5.16); actually, they will propagate along a fan of directions around PQ.

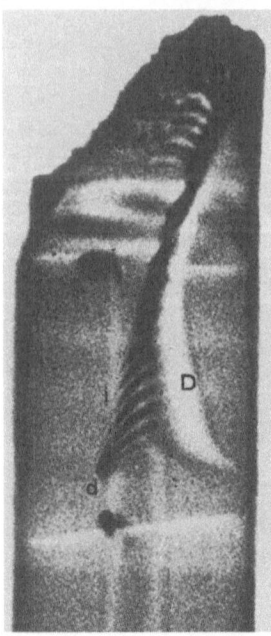

Fig. 5.22. Section topograph of a silicon crystal. (2$\bar{2}$0), Mo K_α. Crystal thickness: 1 mm. D: dynamical image; i: intermediary image; d: direct image

Fig. 5.23. Section topograph of a silicon crystal. (3$\bar{3}\bar{3}$), Mo K_α. Crystal thickness: 1 mm. P: perturbation of the Pendellösung fringes by dislocations in the incidence plane

These new wave fields propagating along PQ are coherent with all the wave fields propagating in the *Borrmann* triangle and will therefore interfere with the wave fields propagating along AQ which have not been influenced by the presence of the dislocation. A set of fringes, characteristic of the intermediary image, will appear which will be superposed on the normal spherical wave pattern. The position of the intermediary image relative to the direct and dynamical image depends on the position of P in the *Borrmann* triangle. The dimension of the intermediary image depends on the position of P and on the angular width of the new wave fields which also depends on the position of P.

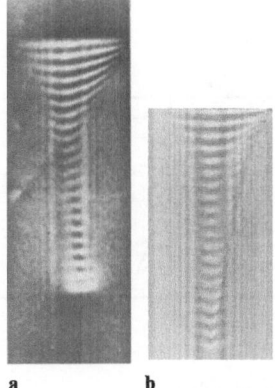

a   b

Fig. 5.24a and b. Dislocation image in *KDP*-(200) reflection, Mo K_α, b=[001] (courtesy of *Dunia* and *Epelboin* [5.104]). (a) Topograph; (b) simulation

Figures 5.19, 5.22 and 5.23 give several examples of intermediary images. It can be seen that the new wave fields can be produced in a fan of aperture up to 2θ.

The intensity of the intermediary image increases as P becomes closer to the direct beam [5.3] and there is no transition between intermediary and direct images.

The intermediary image is well accounted for in the *Taupin-Takagi* theories as has been shown by the excellent agreement between simulations of dislocation images and the corresponding actual section topographs [5.93]. Figure 5.24 shows as example the comparison between simulated and observed dislocation images in *KDP* in a situation of residual contrast ($h \cdot b = 0$).

5.5.5 Deformation of Pendellösung Fringes in the Neighborhood of a Dislocation

We have indicated in Section 5.4.2 that the period of Pendellösung fringes is modified in the presence of strains. This is observed in the case of dislocations and in particular when the dislocations lie parallel to the plane of incidence. *Kato* has first observed and interpreted this effect [5.105]. Figure 5.23 shows an example of the deformation of Pendellösung fringes on each side of the image of a bunch of dislocations lying parallel to the incidence plane. (In this case, the dynamic image of the dislocation lies along the trace of a plane of incidence and is very thin.)

5.6 Images of Planar Defects on Section Topographs

5.6.1 Various Types of Planar Defects

A surface defect is a surface of discontinuity separating two perfect or nearly perfect regions of the cystal. Several different situations may be considered.

1) The two regions are separated by a *gap* [5.106, 107].

2) The two regions are translated with respect to one another through a translation which is *not* a translation of the lattice (*stacking fault, translational*

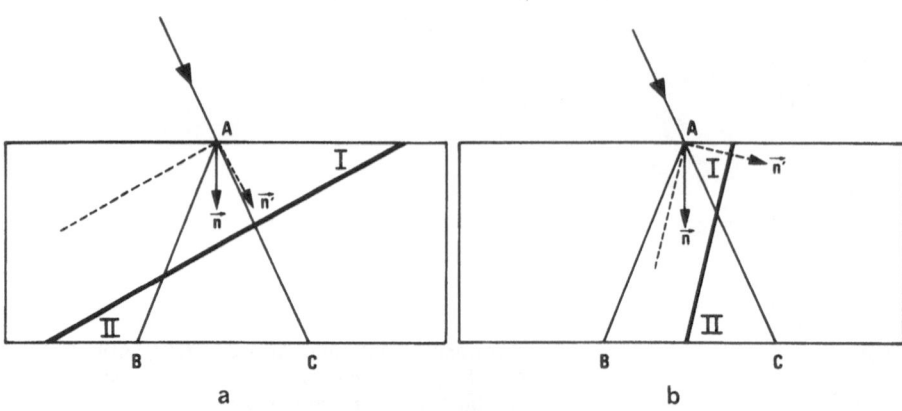

Fig. 5.25a and b. Orientation of a planar defect. (a) *Laue-Laue*; (b) *Laue-Bragg*

twin) or there is a phase difference between the structure factors of the two regions (*twins*) [5.24, 25].

3) The reciprocal lattice vector is different in the two regions: the two lattices are *rotated* with respect to one another and/or have *different spacings* (*misorientation boundary*) [5.20, 26, 27].

4) There is a strain gradient on each side of the boundary as in the case of a layer of micro-inclusions [5.108].

Actually, several of these situations may occur simultaneously.

The calculation of the image contrast depends on the geometrical orientation of the fault. As is shown in Fig. 5.25, it may be either in the *Laue-Laue* or in the *Laue-Bragg* orientation. The former situation only will be described here. The latter has been treated in the case of a stacking fault by *Indenbom* and co-workers [5.109, 110] and discussed by *Schlenker* and co-workers in the case of a wall between ferromagnetic domains in Fe 3% Si [5.111].

There are two main methods to calculate the contrast: to use the *Kato* spherical wave treatment [5.20, 24, 27, 112] or to resolve the *Takagi-Taupin* equations [5.113]. The former provides a physical interpretation for the image formation and will only be considered here.

5.6.2 Stacking Faults

Let us consider the Borrmann triangle and a stacking fault intersecting its sides at B_1 and C_1 (Fig. 5.26). When a wave field reaches the fault at a point such as q, the boundary conditions for the continuity of the tangential component of the wave vectors must be applied along the surface of the stacking fault, just as in the case of a wave incident on a new crystal. Let Q_1 and Q'_2 be the tie-points of two wave fields propagating along Aq (Fig. 5.27). The tie-points excited in Part II of the crystal are at the intersection of the dispersion surface with the normals to the fault surface drawn from Q_1 and Q'_2.

1) Q_1 and Q'_2; we shall call *normal* waves the corresponding wave fields; they propagate along qq' (Fig. 5.26)

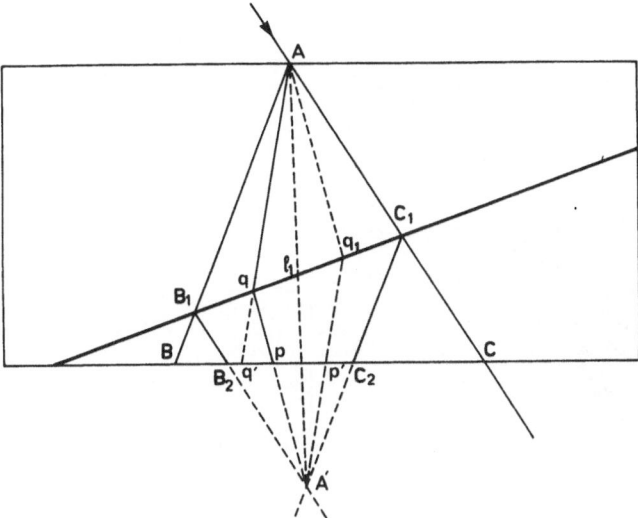

Fig. 5.26. Wave field propagation in a crystal containing a stacking fault. B_1C_1: stacking fault; A': focalization point of the new wave fields

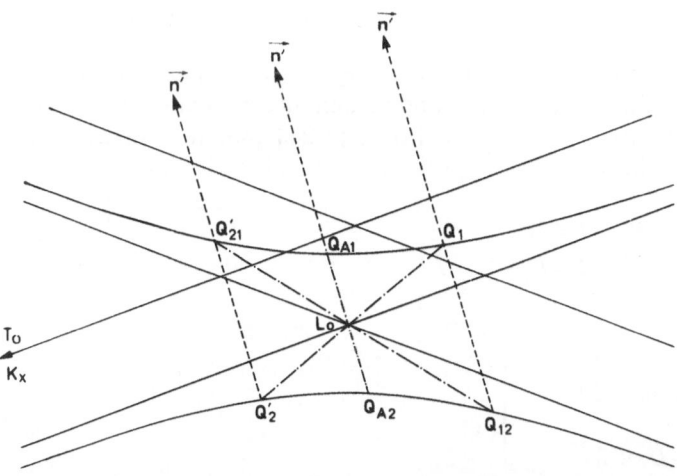

Fig. 5.27. Tie-points of the wave fields at the fault surface. Q_1, Q_2': Normal wave fields; Q_{12}, Q_{21}': new wave fields; Q_{A1}, Q_{A2}: tie-points of the wave fields propagating along AA'; n': normal to the fault surface

2) Q_{12} and Q_{21}'; we shall call *new* waves the corresponding wave fields; they propagate along qp.

It can be shown that the paths of all the new wave fields thus created focalize at a point A', intersection of the parallels to the reflected and incident directions drawn from B_1 and C_1, respectively. If A' lies outside the crystal, it is a *virtual* focal point.

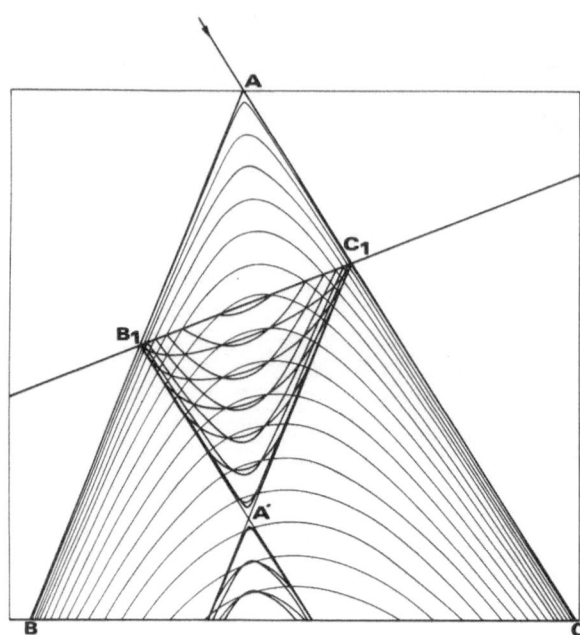

Fig. 5.28. Fringes due to the interference of normal and new wave fields

If one looks at the intensity distribution along the exit surface, one finds that there is a depletion along CC_2 and B_2B and an enhancement along C_2B_2 where the new wave fields arrive. It can be shown [5.24] that the corresponding intensities are of the form

$$
\begin{array}{lll}
\text{along } CC_2 \text{ and } B_2B & I = I_1 \\
\text{along } B_2C_2 & I = I_1 + I_2 + I_3
\end{array}
\tag{5.51}
$$

with

$$
\left.
\begin{array}{l}
I_1 = I_{\text{PER}}(1 - |A| \sin^2 \delta/2) \\
I_2 = B \sin^2 \delta/2 \\
I_3 = C \sin^2 \delta/2 + D \sin \delta
\end{array}
\right\},
\tag{5.52}
$$

where I_{PER} is the intensity for the perfect crystal, and δ the phase shift introduced by the stacking fault

$$
\delta = -2\pi \boldsymbol{h} \cdot \boldsymbol{u}.
\tag{5.53}
$$

$\boldsymbol{u}$ is the displacement vector across the fault and $\boldsymbol{h}$ the diffraction vector.

I_1 is the result of the interference between wave fields which suffered no interbranch scattering when crossing the fault (normal wave fields), that is, wave fields which have propagated along paths such as Ap. It is smaller than the corresponding intensity for the perfect crystal I_{PER} since part of the intensity has

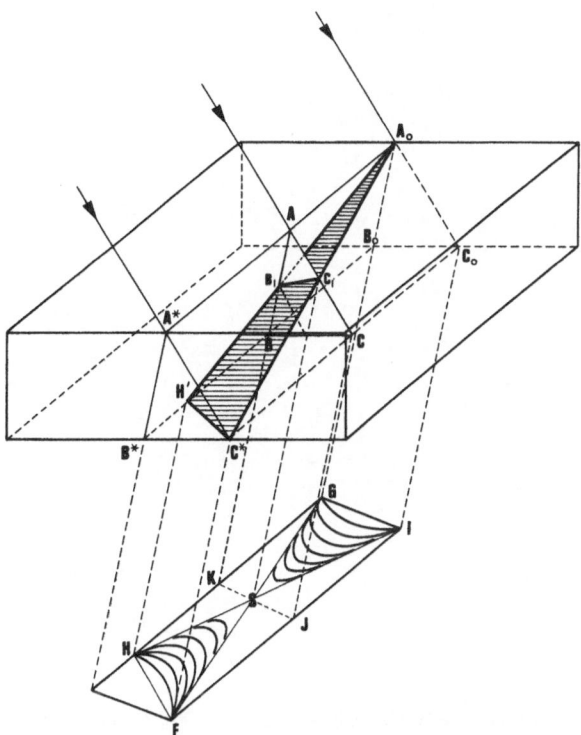

Fig. 5.29. Construction of the image of a stacking fault on a section topograph. *GF*: direct image; *GIS, SHF*: intermediary image; *GSH, ISF*: dynamic image; *HF*: exit surface side; *GI*: entrance surface side

been used for the new wave fields. When μt is not too large, I_1 gives rise to fringes. Equal intensity fringes in the plane of incidence are hyperbolae asymptotic to AB and AC just as in the case of the perfect crystal.

I_2 is the result of the interference between wave fields which have all suffered interbranch scattering (new wave fields), that is, wave fields which have propagated along paths such as Aqp. It also gives rise to fringes when μt is small. Equal intensity fringes are hyperbolae asymptotic to $A'B_1$ and $A'C_1$ (Fig. 5.28).

I_3 is the result of the interference between wave fields which have and have not suffered interbranch scattering, that is, between normal and new wave fields which have followed respectively paths such as Aqp and Ap. The corresponding fringes are the *flat* fringes visible on Fig. 5.28 and passing through intersections of the two sets of preceding hyperbolae. They are visible even for very large values of μt.

Figure 5.29 shows the construction of the I_3 type fringes on the photographic plate when the fault is inclined. The image of the stacking fault takes the now well-known *hourglass* shape. When the focalization point A' lies on the exit surface, all the new wave fields focalize at the same point S of the photographic plate where the intensity is of course very large.

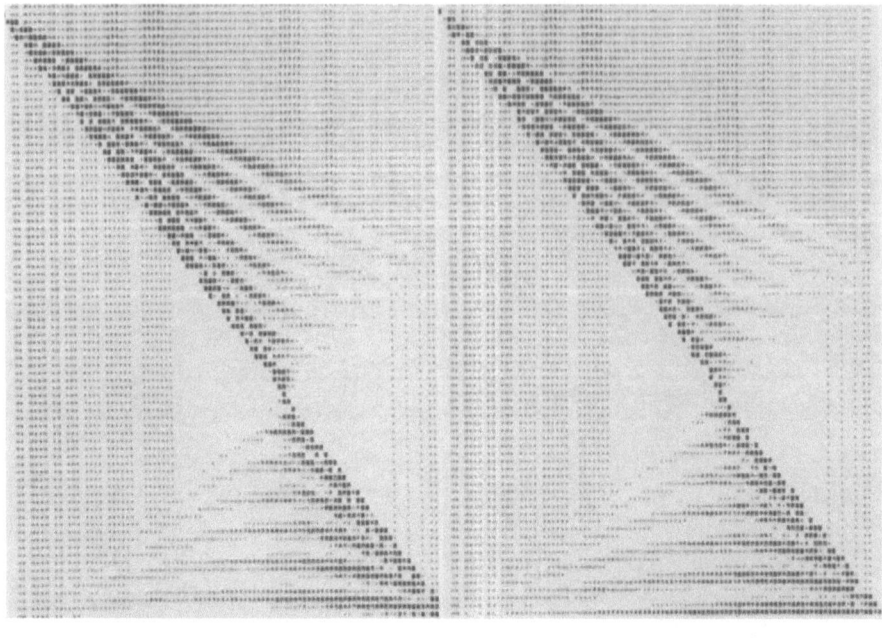

Fig. 5.30a and b. Computer simulations of stacking faults in silicon. Cu K_α; t: 400 μm; crystal surface (111); reflection $\bar{1}11$. (a) fault vector $\frac{1}{3}[\bar{1}\bar{1}1]$; (b) fault vector $\frac{1}{3}[11\bar{1}]$

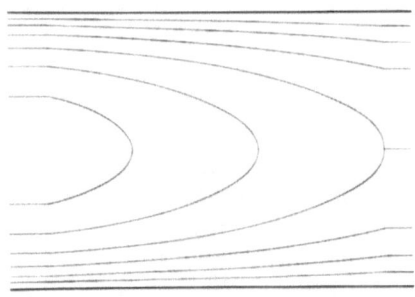

Fig. 5.31. I_1 fringes on the image of a twin boundary on a section pattern

What is observed in practice is the superposition of the three sets of fringes, the I_3 fringes becoming progressively predominant as μt increases [5.114].

It will be noticed that only the second term in I_3 depends on the sign of δ and will enable one to determine the sign of the phase shift.

Figure 5.30 compares computer simulations of extrinsic and intrinsic stacking faults in silicon for which $\delta = \pm 2\pi/3$. It can be seen that the contrast of the first fringe is opposite at the entrance and exit surfaces and for opposite values of δ, enabling an unambiguous determination of the sign of δ. The

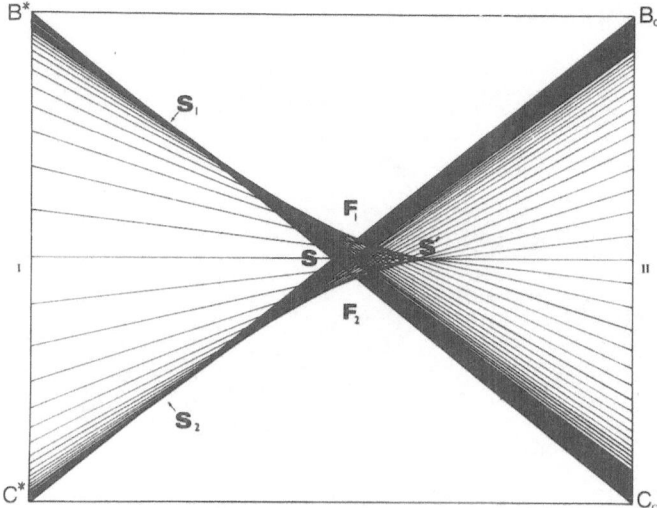

Fig. 5.32. Localization of the new wave fields in the image of a twin boundary on a section pattern

contrast of the first fringes can be directly derived from the approximate value of D when the fault is very close either to the entrance or to the exit surface.

$$D = \pm |A_0|^2 \pi z_j \sinh 2\Phi_i / t\Lambda (1 \mp Y) \qquad (5.54)$$

where the top sign corresponds to the entrance surface, z_j is the distance of the fault from the relevant surface, t is the crystal thickness, and Φ_i is related to the diffraction parameters and proportional to the absorption coefficient. Equation (5.54) is valid only in a symmetric case.

The contrast varies with the value of δ, and by comparing an actual topograph with simulations performed with various values of δ it is possible to determine the phase shift [5.115].

When the stacking fault is enclosed by a partial dislocation loop and does not intersect any of the crystal surfaces, it is not possible to determine the contrast of the first fringe and therefore the nature of the fault. From the shape of the image it is however possible to determine accurately its depth in the crystal [5.116].

When traversing crystal and photographic plate to obtain a projection topograph, the image of the stacking fault is integrated along a direction parallel to its intersection with the exit surface. A set of parallel fringes is thus obtained. The contrast of the first fringe is the same as on the section pattern for a symmetrical case but not necessarily so in the general case [5.25]. Section patterns are therefore the best tool for the exact characterization of a stacking fault.

5.6.3 Twin Boundaries

We shall consider the case where the structure factor has a different absolute value on each side of the fault. The values of the Fourier coefficients of the

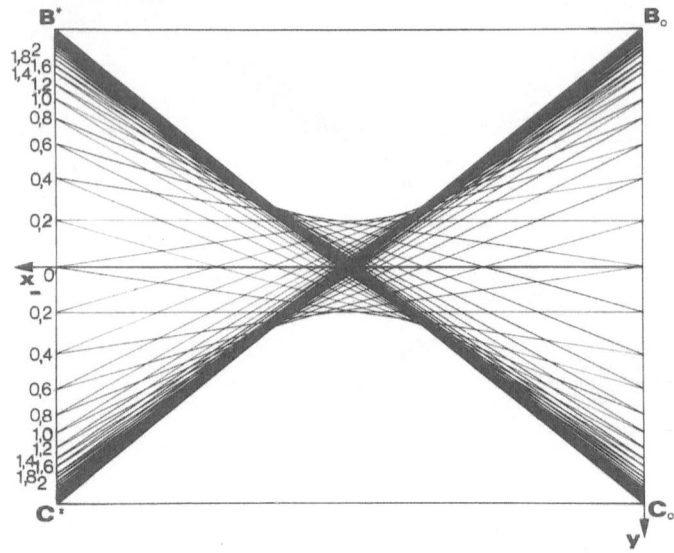

Fig. 5.33. Localization of the new wave fields in the image of a misorientation boundary on a section pattern

electric susceptibility are therefore also different: the diameter of the dispersion surface is different and the period of the *Kato* spherical-wave Pendellösung fringes is also different.

The intensity distribution can still be considered as the sum of three terms I_1, I_2, I_3 having the same interpretation, but their expressions are much more complicated. The I_1 term corresponds to the adaptation between the two sets of Pendellösung fringes on each side of the fault (see Fig. 5.31). The domain of definition of the new wave fields is now limited by a caustic and there is no focalization any more (Fig. 5.32). Computer simulations of the image have been performed by *Kato* and co-workers [5.20, 26].

5.6.4 Misorientation Boundaries

We shall assume here that the two parts of the crystal on each side of the boundary are perfect, have the same structure factor but are slightly misoriented respective to one another, and/or have a slightly different lattice parameter. They have therefore different reciprocal lattice vectors h^I and h^{II}. The effective misorientation between the two parts is equal to

$$\delta\Theta = s_h \cdot (h^{II} - h^I)/k \sin \Theta . \tag{5.55}$$

When a wave field reaches the boundary, the boundary conditions for the wave vectors have to be applied as in the case of a stacking fault. But since the reciprocal lattice vector is different in Part II, the dispersion surface correspond-

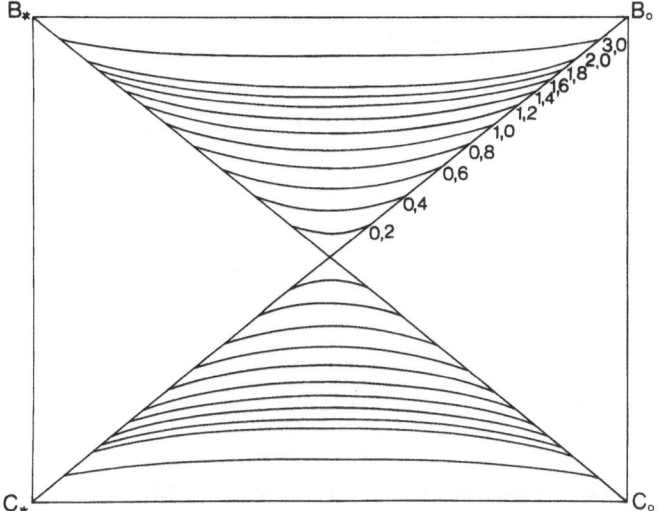

Fig. 5.34. Variation of the shape of the image of a misorientation boundary on a section topograph with the value of the misorientation. The numbers correspond to $\delta\theta/\delta$ where $\delta\theta$ is the misorientation and δ the linewidth for the perfect crystal

ing to Part II is shifted by the quantity $k\delta\theta$ along the asymptote T_0 (see Fig. 5.7). This complicates matters and there is no more focalization of the new wave fields (Fig. 5.33). It is possible to relate the minimum width δy of the domain occupied by the new wave fields on the section pattern to the effective misorientation $\delta\Theta$ between the two parts of the crystal

$$\delta y = B_0 C_0 \delta\eta / \sqrt{4 + \delta\eta^2} \tag{5.56}$$

where $\delta\eta = \delta\Theta \sin 2\Theta / \sqrt{\chi_h \chi_{\bar{h}}}$ and $B_0 C_0$ is the total width of the section pattern.

Figure 5.34 shows how the pattern varies with the value of $\delta\eta$ and Fig. 5.35 shows as example the image on a section topograph of a boundary between two neighboring growth sectors in a potash alum crystal. It can be seen that it is characteristic of a misorientation boundary. From the width of the image in its center it can be found that the effective misorientations between the two growth sectors is of about 0.45 s of arc. Computer simulations of the image of a misorientation boundary on a section topograph have been performed by *Kato* and co-workers [5.20, 26].

5.6.5 Boundaries Associated with a Strain Gradient

Let us consider a distribution of strain sources along a surface. It might be, for instance, a distribution of micro-inclusions decorating a growth band. These micro-inclusions induce strains in the crystal which decrease with the distance from the fault.

Fig. 5.35. X-ray topograph of an alum crystal. Mo K_α.(220) reflection. Crystal thickness: 1.5 mm (courtesy of *Robert*, private communication)

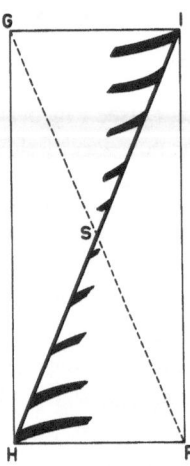

Fig. 5.36. Schematic diagram of the image of a boundary associated with a strain gradient

If the strain gradient is large enough close to the fault, interbranch scattering will occur exactly as in the vicinity of a dislocation and new wave fields are expected to be generated. Since the fault is plane, one expects as a first approximation that the localization of these wave fields will be similar to that occurring in a stacking fault. Furthermore, the intensity of this *intermediary* image will be highest close to the direct image of the fault. The expected image is schematically represented in Fig. 5.36 [5.108]. Such a pattern can be observed on Fig. 5.2b which represents the image of growth band in quartz on a section topograph. A second phenomenon is superposed on the first one in this case. There is a relaxation of the stresses associated with this growth band at its intersection with the crystal surfaces and this gives rise to long range strains in the crystal. These long range strains induce new fringes in the *Kato* spherical-wave Pendellösung fringe system as described in Section 5.4.2. These new fringes can be seen in Fig. 5.2b around the direct images of the intersections of the growth band with the two crystal surfaces

References

5.1   W.Berg: Z. Naturwiss. **19**, 391 (1931)
5.2   U.Bonse, M.Hart, J.B.Newkirk: Advan. X-Ray Anal. **10**, 1 (1967)
5.3   A.Authier: Advan. X-Ray Anal. **10**, 9 (1967)
5.4   A.R.Lang: Advan. X-Ray Anal. **10**, 91 (1967)
5.5   A.R.Lang: In *Modern Diffraction and Imaging Techniques in material Science*, ed. by S. Amelinckx, R. Gevers, G. Remaut, J. van Landuyt (North Holland, Amsterdam 1970) p. 407
5.6   G.Hildebrandt: Fortschr. Mineral. Dtsch. **53**, 1 (1975)
5.7   B.K.Tanner: *X-Ray Diffraction Topography* (Pergamon Press, Oxford 1976)
5.8   J.B.Newkirk: J. Appl. Phys. **29**, 995 (1958)
5.9   J.B.Newkirk: Phys. Rev. **110**, 1465 (1958)
5.10 A.P.L.Turner, T.Vreeland Jr., D.P.Pope: Acta Cryst. **A24**, 452 (1968)
5.11 A.R.Lang: J. Appl. Phys. **29**, 597 (1958)
5.12 A.R.Lang: Acta Cryst. **12**, 249 (1958)
5.13 A.R.Lang: J. Appl. Phys. **30**, 1748 (1959)
5.14 K.Haruta: J. Appl. Phys. **36**, 1789 (1965)
5.15 T.Vreeland Jr.: J. Appl. Cryst. **9**, 34 (1976)
5.16 A.R.Lang: Acta Met. **5**, 358 (1957)
5.17 N.Kato, A.R.Lang: Acta Cryst. **12**, 787 (1959)
5.18 N.Kato: Acta Cryst. **14**, 526, 627 (1961)
5.19 N.Kato: J. Appl. Phys. **39**, 2225, 2231 (1968)
5.20 N.Kato: In *X-Ray Diffraction*, ed. by L.V. Azaroff, R. Kaplov, N. Kato, R.J.Weiss, A.J.C. Wilson, R.A. Young (McGraw-Hill, New York 1974) p. 222
5.21 N.Kato: In *Crystal Growth and Characterization*, ed. by R. Ueda, J.B. Mullin (North Holland, Amsterdam 1975) p. 279
5.22 A.Authier: In *Modern Diffraction and Imaging Techniques in Material Science*, ed. by S. Amelinckx, R. Gevers, G. Remaut, J. van Landuyt (North Holland, Amsterdam 1970) p. 481
5.23 J.R.Patel: J. Appl. Phys. **44**, 3903 (1973)
5.24 A.Authier: Phys. Stat. Sol. **27**, 77 (1968)
5.25 A.Authier, J.R.Patel: Phys. Stat. Sol. **27a**, 213 (1975)
5.26 T.Katagawa, H.Ishikawa, N.Kato: Acta Cryst. **A31**, 246 (1975)
5.27 A.Authier, M.Sauvage: J. Physique **27**, C3–137 (1966)
5.28 A.Authier, M.C.Robert, A.Zarka: (to be published)
5.29 U.Bonse, E.Kappler: Z. Naturforsch. **13a**, 348 (1958)
5.30 U.Bonse: Z. Physik **153**, 278 (1958)
5.31 M.Renninger: Phys. Letters **1**, 104 (1962)
5.32 M.Renninger: Z. Naturforsch. **199**, 783 (1964)
5.33 M.Renninger: Z. Angew. Phys. **19**, 20 (1965)
5.34 K.Kohra, H.Hashizume, J.I.Yoshimura: Jap. J. Appl. Phys. **9**, 1029 (1970)
5.35 M.Hart: Sci. Progr. (London) **56**, 429 (1968)
5.36 A.R.Lang: Brit. J. Appl. Phys. **14**, 904 (1963)
5.37 J.L.Caslavsky: Rev. Sci. Inst. **41**, 517 (1970)
5.38 J.I.Chikawa, Y.Asaeda, I.Fujimoto: J. Appl. Phys. **41**, 1922 (1970)
5.39 U.Bonse: In *Direct Observation of Imperfections in Crystals*, ed. by J.B.Newkirk, J.H.Wernick (J. Wiley, New York 1962) p. 431

5.40   J.F.Petroff, A.Authier: Phys. Stat. Sol. 13, 373 (1966)
5.41   M.Renninger: J. Appl. Cryst. 9, 178 (1976)
5.42   M.Englander, C.Malgrange, J.F.Petroff, M.Sauvage, A.Zarka: Phil. Mag. 33, (5), 743 (1976)
5.43   H.Tomimitsu, K.Doi: J. Appl. Cryst. 7, 59 (1974)
5.44   M.Schlenker, J.Baruchel, R.Perrier de la Bathie, S.H.Wilson: J. Appl. Phys. 46, 2845 (1975)
5.45   M.Ando, S.Hosoya: Phys. Rev. Letters 29, 281 (1972)
5.46   J.B.Davidson, S.A.Werner, A.S.Arrott: AIP Conf. Proc. 18, 396 (1974)
5.47   J.B.Davidson: J. Appl. Cryst. 7, 356 (1974)
5.48   M.Schlenker, C.G.Shull: J. Appl. Phys. 44, 4181 (1973)
5.49   J.Baruchel, M.Schlenker, W.L.Roth: J. Appl. Phys. 48, 5 (1977)
5.50   C.Darwin: Phil. Mag. 27, 315, 675 (1914)
5.51   P.P.Ewald: Ann. Phys. Dtsch. 54, 519 (1917)
5.52   M. Laue, W.Friedrich, P. Knipping: München Sitzungsbericht 383 (1912)
5.53   G.Borrmann: Z. Physik 42, 157 (1941)
5.54   G.Borrmann: Z. Physik 127, 297 (1950)
5.55   H.Campbell: Acta Cryst. 4, 180 (1951)
5.56   M. von Laue: Acta Cryst. 2, 106 (1949)
5.57   G.Borrmann: Z. Krist. 106, 109 (1954)
5.58   M. von Laue: Acta Cryst. 5, 619 (1952)
5.59   G.Borrmann: Z. Naturwiss. 42, 67 (1955)
5.60   H.Wagner: Z. Physik 146, 127 (1956)
5.61   N.Kato: Acta Cryst. 11, 885 (1958)
5.62   P.P.Ewald: Acta Cryst. 11, 888 (1958)
5.63   E.H.Wagner: Z. Physik 154, 352 (1959)
5.64   G.Borrmann: Beiträge zur Physik und Chemie des 20. Jahrhunderts (Vieweg Braunschweig 1959) p. 262
5.65   N.Kato: Acta Cryst. 13, 349 (1960)
5.66   S.Takagi: Acta Cryst. 15, 1311 (1962)
5.67   M. von Laue: Röntgenstrahlinterferenzen (Akademie Verlagsgesellschaft, Frankfurt am Main 1960)
5.68   A.Authier: Advan. Struct. Res. Diffraction Methods 3, 1 (1970)
5.69   B.Borie: Acta Cryst. 23, 210 (1967)
5.70   M. von Laue: Ergeb. Exact. Naturw. 10, 133 (1931)
5.71   S.Takagi: J. Phys. Soc. Jap. 26, 1239 (1969)
5.72   D.Taupin: Bull. Soc. Franc. Mineral. Crist. 87, 469 (1964)
5.73   A.Authier: J. Phys. Radium 27, 57 (1966)
5.74   G.Borrmann, G.Hildebrandt, H.Wagner: Z. Physik 142, 406 (1955)
5.75   A.Authier, F.Balibar: Acta Cryst. A26, 647 (1970)
5.76   P.Penning, D.Polder: Philips Res. Rept. 16, 419 (1961)
5.77   N.Kato: J. Phys. Soc. Jap. 18, 1785 (1963)
5.78   N.Kato: J. Phys. Soc. Jap. 19, 67, 971 (1964)
5.79   Y.Ando, N.Kato: Acta Cryst. 21, 284 (1966)
5.80   M.Hart: Z. Physik 189, 269 (1966)
5.81   N.Kato, J.R.Patel: J. Appl. Phys. 44, 965 (1973)
5.82   J.R.Patel, N.Kato: J. Appl. Phys. 44, 971 (1973)
5.83   Y.Ando, J.R.Patel, N.Kato: J. Appl. Phys. 44, 4405 (1973)
5.84   Y.Ando, N.Kato: J. Appl. Cryst. 3, 74 (1970)
5.85   A.Authier, C.Malgrange, M.Tournarie: Acta Cryst. A24, 126 (1968)
5.86   T.Katagawa, N.Kato: Acta Cryst. A30, 830 (1974)
5.87   P.Penning: Theory of X-Ray diffraction in unstrained and lightly strained perfect crystals (Thesis, Univ. of Delft 1966)
5.88   F.Balibar: Acta Cryst. A24, 666 (1968)
5.89   F.Balibar: Acta Cryst. A25, 650 (1969)
5.90   F.Balibar, C.Malgrange: Acta Cryst. A31, 425 (1975)

5.91   F.Balibar, Y.Epelboin, C.Malgrange: Acta Cryst. A31, 836 (1975)
5.92   A.Authier, F.Balibar, Y.Epelboin: Phys. Stat. Sol. 41, 225 (1970)
5.93   F.Balibar, A.Authier: Phys. Stat. Sol. 21, 413 (1967)
5.94   K.Kambe: Z. Naturforsch. 18a, 1010 (1963)
5.95   G.Borrmann: Physik Bl. Dtsch. 15, 508 (1959)
5.96   A.Authier: Bull. Soc. Franc. Mineral. Crist. 84, 115 (1961)
5.97   Z.Ishii: J. Phys. Soc. Jap. 17, 838 (1962)
5.98   A.R.Lang, M.Polcarova: Proc. Roy. Soc. A285, 297 (1965)
5.99   M.Wilkens, F.Meier: Z. Naturforsch. 18a, 26 (1963)
5.100 J.I.Chikawa: J. Appl. Phys. 36, 3496 (1965)
5.101 C.Willaime, A.Authier: Bull. Soc. Franc. Mineral. Crist. 89, 269 (1966)
5.102 J.E.A.Miltat, D.K.Bowen: J. Appl. Cryst. 8, 657 (1975)
5.103 H.Klapper: J. Appl. Cryst. 9, 310 (1976)
5.104 E.Dunia, Y.Epelboin, J.F.Petroff: (to be published)
5.105 N.Kato: In *Crystallography and Crystal Perfection*, ed. by G.N. Ramachandran (Academic Press, London 1963) p.153
5.106 A.Authier, D.Milne, M.Sauvage: Phys. Stat. Sol. 26, 469 (1968)
5.107 S.Tannemura, A.R.Lang: Z. Naturforsch. 28a, 668 (1973)
5.108 A.Authier, M.C.Robert, A.Zarka: (to be published)
5.109 V.L.Indenbom, F.N.Chukhovskii: Sov. Phys. Cryst. 19, 19 (1974)
5.110 V.L.Indenbom, I.Sh.Slobodetskii: Soc. Phys. Cryst. 19, 23 (1974)
5.111 M.Schlenker, P.Brissonneau, J.Perrier: Bull. Soc. Franc. Mineral. Crist. 91, 653 (1968)
5.112 N.Kato, K.Usami, T.Katagawa: Advan. X-Ray Anal. 10, 46 (1967)
5.113 A.Authier, D.Simon: Acta Cryst. A24, 517 (1968)
5.114 B.C.Wonsiewicz, J.R.Patel: J. Appl. Phys. 47, 1837 (1976)
5.115 A.Authier, Y.Epelboin: Phys. Stat. Sol.(a) 41, K9 (1977)
5.116 J.R.Patel, A.Authier: J. Appl. Phys. 46, 118 (1975)
5.117 J.F.Petroff, A.Authier: Phys. Stat. Sol. 17, K3 (1966)

6. Live Topography

W. Hartmann

With 25 Figures

X-ray topography is a widely applied technique for a nondestructive analysis of the crystalline perfection of solids. All topography methods utilize the fine structure within Laue diffraction spots, which is generated by deviations from perfection of the lattice. Topography consists in imaging the defect structure with the light of one particular Laue reflex. Figure 6.1 is a schematic drawing of an X-ray topography arrangement. An X-ray beam irradiates a crystal which is set in Bragg condition for one wavelength and definite (*hkl*) lattice planes. All X-rays which are diffracted from the lattice planes in Bragg condition are superimposed and produce the diffracted beam. An X-ray topogram of a crystal area without lattice defects shows a uniformly grey picture on a photographic plate. This picture has the same dimensions as the irradiated crystal area. Contrasts appear if lattice defects, for example dislocations, are in the imaged crystal area.

Topography techniques are indispensable and standard methods to determine the defect structure, excluding point defects, of crystal lattices, since the first investigations in the X-ray topography field done by *Berg*, 1931 [6.1]; *Barrett*, 1945 [6.2]; *Lang*, 1958 [6.3, 4]; *Barth* and *Hosemann*, 1958 [6.5]; and *Newkirk*, 1959 [6.6]. The first section in this chapter will give a short introduction of the most commonly employed techniques.

The spatial resolution of the standard methods is of the order of several μm, but the time resolution is poor, since several hours are typically needed to obtain a complete image of the crystal and prepare the final image document.

A live topography method should have a similar spatial resolution of several μm to resolve the fine structure of a Laue spot and simultaneously a time resolution of fractions of a second to obtain a live observation of the defect configuration. A live topography method needs an electro-optical device, for example a television (TV) system, to image the X-ray topograms and to obtain the time resolution. A TV system can usefully employ the advantages of electronic image processing. The superposition of several TV frames, for example, increases the visibility of defects, but decreases the time resolution. This image processing mode indicates an exchange of time and spatial resolution and can be made either analog or digital. A subtraction of successive TV frames can be obtained only by a digital image processing. Only changes of defect contrasts are seen if TV frames are subtracted.

Much development was made since 1961 to increase the time resolution, because of the obvious interest in direct studies of the variation of the defect

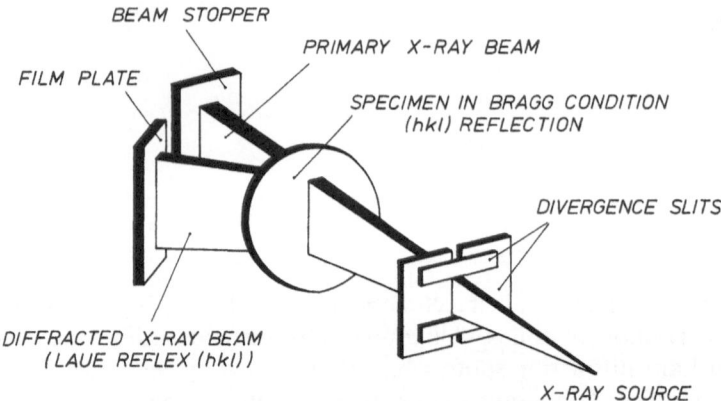

BEAM STOPPER

FILM PLATE

PRIMARY X-RAY BEAM

SPECIMEN IN BRAGG CONDITION
(hkl) REFLECTION

DIVERGENCE SLITS

DIFFRACTED X-RAY BEAM
(LAUE REFLEX (hkl))

X-RAY SOURCE

Fig. 6.1. Schematic illustration of an X-ray topographic arrangement in transmission geometry

structure on a time scale during which influences act on the crystal. The recent efforts with good time and spatial resolution are discussed in Section 6.2. Next we give a comparison between these methods and present limitations. Finally some future trends and possibilities are indicated.

6.1 X-Ray Topography

X-ray topography nondestructively identifies dislocations, stacking faults, grain boundaries, and other, one- or more-dimensional lattice imperfections. These imperfections are imaged by suitable selected reflections, as indicated by their Miller indices (hkl). The amount and direction of the Burgers vector of dislocations can be determined by accomplishment of the contrast criteria. The Burgers vector determines the nature of the defect. There is no contrast if the lattice vector of the reflecting net planes is normal to the Burgers vector. Maximum contrast exists if the lattice vector is parallel to the Burgers vector. A more detailed investigation of the visibility of lattice defects and contrast phenomena is reported by *Lang* [6.10] and *Authier* [6.11]. The various standard topography techniques are

1) Berg-Barrett method,
2) Double-crystal methods,
3) Lang method.

The Berg-Barrett method is most adaptable to topography of shallow surface regions. The geometry of this method is readily accomplished with the longer X-ray wavelengths (Cu or Cr, K_α) to obtain good contrast. The conditions for good resolution and image quality are described by *Austerman* and *Newkirk* [6.7]. Double-crystal methods are very powerful to detect small lattice strains because the angular and wavelength spreads can be reduced very well in a wide range. These methods are applicable in reflection and transmission geometry.

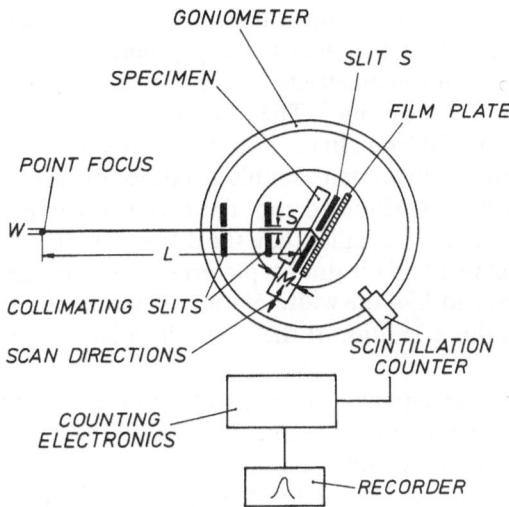

Fig. 6.2. Schematic diagram of an X-ray topographic setup using the Lang method

The properties and principles of the double-crystal arrangements are described by *Kohra* et al. [6.8, 9]. The most widely used technique is the Lang method. A short discussion about its geometry and its principles will be given now. (More details of image quality, resolution, and contrast information are reported by *Lang* [6.10], *Authier* [6.11], and *Tanner* [6.12].)

Figure 6.2 is a schematic diagram of a topography arrangement using the Lang method. An X-ray beam enters the specimen at a large distance from the X-ray source. The beam divergence of about 5×10^{-4} rad is produced by the collimator slits. The specimen is set in Bragg condition, usually for K_{α_1} radiation and a definite (*hkl*) reflection with the help of a scintillation counter. The slit S behind the sample stops the primary beam. Only the diffracted beam which contains the information about lattice defects reaches the fine grain photographic plate. A picture of the entire crystal is obtained by traversing the crystal and the film together across the X-ray beam. The photographic plate is developed after an exposure time of several minutes up to a few hours. The exposure time depends on the power of the X-ray generator, the crystal thickness, the crystal size and material, the used (*hkl*) reflection, and the X-ray wavelength. The contrasts of defects in the crystal are seen in a 1:1 imaging after the developing procedure.

The geometry of the arrangement is important to achieve a good spatial resolution. The resolution in the vertical direction is determined only by geometrical factors and given by the expression

$$D_v = \frac{M}{L} H \tag{6.1}$$

where D_v is the distance of two points which can be resolved, M is the distance between crystal and photographic plate, L is the distance from X-ray source to the crystal (see Fig. 6.2), and H is the focus size in vertical direction. The vertical focus size H is about 0.1 mm in standard fine focus X-ray generators. The distance M is limited to about 10 mm for geometrical reasons to separate the primary and secondary beams behind the specimen. The resolution D_v becomes about 1.5 μm with a distance L of about 700 mm. In the horizontal direction, K_{α_1} radiation should be diffracted only. Unwanted double pictures of defect contrasts appear, if K_{α_2} radiation has been diffracted, too. The divergence of the X-ray beam defined by $(W + L_s)/L$ should be smaller than the difference $\Delta\Theta_{1-2}$ of the Bragg angles of the K_α doublet to avoid the double pictures. The focus size in horizontal direction is denoted W, and L_s is the width of the collimator slits in Fig. 6.2. It is no problem to fulfill this condition if the beam divergence is of about 5×10^{-4} rad.

All photographic methods, while capable of excellent resolution and contrast, are generally inapplicable to dynamic experiments in which the defect configuration varies continuously with time under external perturbations such as pressure or temperature gradients. There is a definite need for methods of direct observation. Dislocation mobilities have been measured by various techniques such as etching [6.13], X-ray topography [6.14–16], and electron microscopy [6.17]. The etching and topography techniques did not involve continuous observation of dislocation motions. Dislocation velocities and formations were obtained by the comparison of dislocation configurations before and after the application of a stress pulse and/or a temperature variation. Transmission electron microscopy has been the only technique to observe moving dislocations in situ. Electron microscopy observations suffer, however, from one distinct disadvantage. Extremely thin specimens are required. Their preparation may result in a situation different from the dislocation behavior in the bulk state.

It is necessary to have a direct-display topography technique for in situ observations for these reasons. The important details of individual lattice defects—such as dislocations in silicon—are of the order of a few μm. They demand a technique of similar spatial resolution. Direct-display methods, therefore, require a combination of large X-ray intensity at a small focus size with high resolution and good signal-to-noise ratio (SNR) at the detector system. A system fulfilling these conditions would be extremely useful for direct and nondestructive inspection of solid-state technology processes. Such a system is also a necessary requirement for in situ observations of such basic phenomena as creation and motion of dislocations under external stresses, temperature gradients, or during crystal growth.

6.2 Direct-Display Systems

Several investigators [6.18–26] tried from 1961 to 1968 to visualize X-ray diffraction patterns with TV systems. All these systems, however, could not resolve the fine structure of a Laue spot. Direct-display X-ray topography of single crystals was thus impossible. The first system which overcame the difficulties in resolution and sensitivity to resolve the contrasts of individual dislocations was developed by *Chikawa* and *Fujimoto* [6.27] in 1968. They used a rotating anode X-ray generator (see Chap. 2 of this volume) which can be operated at 60 kV and 500 mA with a focus size of $0.5 \times 10\,\text{mm}^2$, the geometry of the Lang method, and a lead oxide vidicon as detector. The spatial resolution was about 30 μm. Seven electro-optical systems were reported from 1969 to 1970. *Chester* et al. [6.28] and *Chester* and *Koch* [6.29] used a silicon diode array camera tube to image diffracted X-rays. *Meieran* et al. [6.30] employed a fluorescent screen to convert X-rays into visible light. The optical pattern was then imaged on a single-stage image intensifier by a high aperture lens. The image intensifier was fiber-optically coupled to a secondary electron conduction (*SEC*) vidicon tube. *Lang* and *Reifsnider* [6.31] reported a system which coupled the optical pattern on a fluorescent screen to a four-stage image intensifier tube by a lens. This image intensifier has a sensitivity such that "10 X-ray photons $\text{s}^{-1}\,\text{mm}^{-2}$ were clearly seen on the intensifier output phosphor by the naked eye". The image on the output of the intensifier tube was lens coupled to a conventional vidicon for display on the TV monitor. *Blamoutier* [6.32] reported about two X-ray sensitive vidicons. The photoconductor used in these tubes consisted of a mixture of selenium, arsenic, and tellurium. *Driard* [6.33] constructed two two-stage image intensifier tubes for application in the Lang topography technique. Each of these tubes possessed a fluorescent screen sandwiched with the photocathode inside the image tube just behind the input beryllium faceplate. Unfortunately, no X-ray topograms were shown in the last two papers. The system of *Ball* and *Furnas* [6.34] was similar to that reported by *Meieran* et al. [6.30]. It employed a SEC vidicon camera tube fiber-optically coupled to the fiber optic faceplate of an image intensifier. The fluorescent screen was in direct contact with the external surface of the fiber optic. All these systems, briefly described before, had a spatial resolution lower than 30 μm and have been discussed in more detail in the review article by *Green* [6.35]. A number of additional papers have appeared since 1971 in which electro-optical systems have been used to look at X-ray topographic images. A survey about these systems is given by *Green* [6.36]. The most recent paper was published by *Eckers* and *Oppolzer* [6.37]. The detector system employed consists of an X-ray sensitive vidicon with an amorphous selenium target developed by Heimann GmbH.

All reported methods may be grouped into two broad categories depending on the general principle used to permit rapid viewing and recording of X-ray topographic images, namely the single-stage imaging and the multiple-stage imaging methods. In the single-stage imaging method an X-ray sensitive vidicon tube directly converts the X-ray topographic image into an electronic charge pattern. This charge pattern is read out by a scanning electron beam and displayed as a visible image on a television monitor. In the multiple-stage imaging method the X-ray topographic image is converted into a visible light pattern by a fluorescent screen. The visible light pattern is then optically coupled

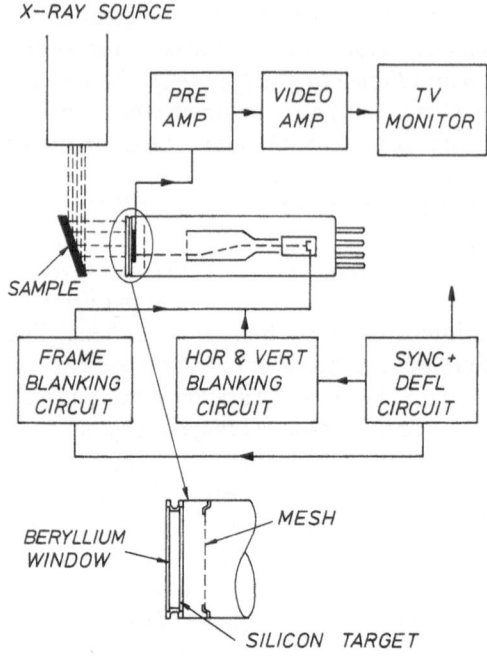

X−RAY SOURCE

Fig. 6.3. Schematic diagram of the system used by *Rozgonyi* et al.

either by a lens or a fiber-optic plate to the input photocathode of a light sensitive electro-optical device. The output image may either be viewed directly or displayed on a TV monitor.

It is necessary to have simultaneously high spatial and high time resolution to observe individual dislocations or other lattice defects and their interaction during dynamical experiments. The chances of meeting this difficult requirement will now be discussed for representative systems. Developments which obtain the X-ray topogram by single-stage imaging are described in Section 6.2.1. Methods which operate with multiple-stage imaging are discussed in Section 6.2.2.

6.2.1 Single-Stage Imaging

Rozgonyi et al. [6.38] reported a video display system with a spatial resolution better than 15 μm in 1970.

Imaging System

The experimental arrangement was that of the Berg-Barrett method. Figure 6.3 is a schematic diagram of the system, and Fig. 6.4 shows the real experimental setup. The X-ray generator was General Electric XRD-5 with a type CA-8S spot focus tube. Chromium K_α radiation was used at 45 kV and 22 mA. The TV camera tube possessed a 100 μm thick beryllium faceplate. A 15–20 μm thick silicon target containing more than 750 000 diodes was placed behind the faceplate at a distance of 250 μm. The target thickness gave an absorption of

Fig. 6.4. Experimental setup of the system of *Rozgonyi* et al.

about 40 % of Cr K_α radiation. The camera was used in a 525-line closed circuit video system. A blanking circuit allowed the target to integrate the charge pattern on the silicon target before it was read out by the electron beam. The active surface of the tube target was 12.7 mm × 9.78 mm and the electronic magnification 30–60 times.

Experiments

A Berg-Barrett topogram of a silicon integrated circuit (IC) pattern in (311) reflection is reproduced in Fig. 6.5. Figures 6.5a and 6.5b are optical photographs of an IC element with an isolated 13 µm metallized area and interlaced 38 µm strips with 19 µm separation. Figure 6.5c is topogram taken with Kodak dental film. Figure 6.5d shows the same area but was photographed from the video monitor. Arrow A indicates the isolated metallized area and arrow B the interlaced strips. The image is displayed every 1/5 s on a TV monitor at this magnification. Figure 6.6b shows an optical micrograph of another IC in higher magnification than in Fig. 6.5 [6.39]. Arrow A indicates a metallized strip of 10 µm width. The same area is shown in Fig. 6.6a taken with the video monitor. The result of a third experiment [6.39] is reproduced in Fig. 6.7. A boron diffusion through oxide strips of various width was made in silicon. Figure 6.7a is an optical photograph and Fig. 6.7b the correlated video image of the crystal taken also in the Berg-Barrett method. The scanned target height should be reduced to 7.9 mm, and the electron spot size must be no larger than 15 πm [6.35]

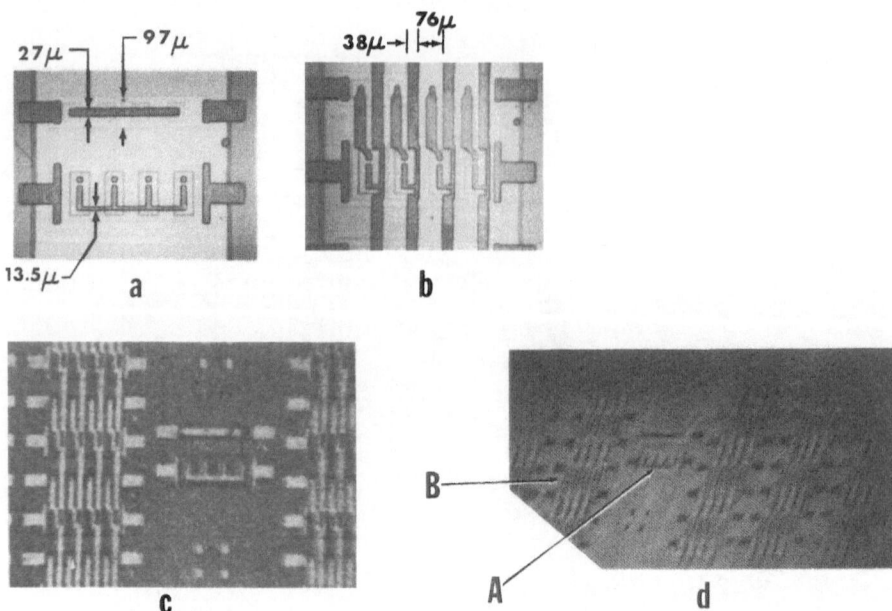

Fig. 6.5. (a) and (b) optical images of a silicon integrated circuit with typical dimensions; (c) X-ray topogram taken with Kodak dental film; and (d) video X-ray topogram (after *Rozgonyi* et al.)

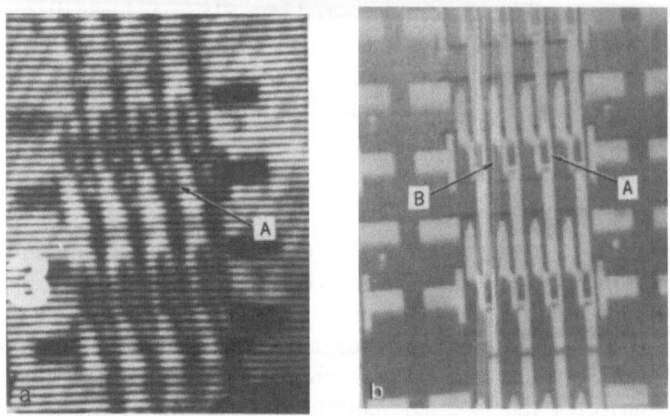

Fig. 6.6a and b. Pictures which indicate the highest resolution of the system of *Rozgonyi* et al.; (a) video X-ray topogram; (b) optical micrograph of a silicon integrated circuit. Arrow A indicates a metallized strip of 10 µm width

if the video system used by *Rozgonyi* et al. should achieve a spatial resolution of 15 µm. The resolution capabilities seem to be attainable because the absorption of the radiation used was about 40 % and the contrasts of metallized areas are high in the reflection technique. Therefore, the SNR of the video signal should be sufficient to obtain the resolution of the TV system of about 15–20 µm. This

WIDTH OF OXIDE CUT

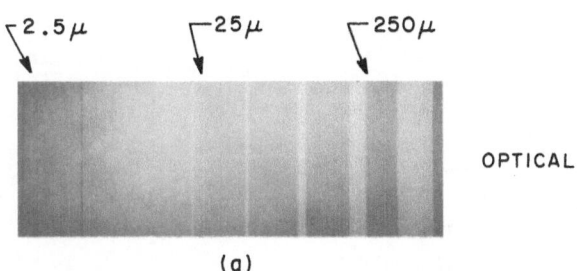

(a)

OPTICAL

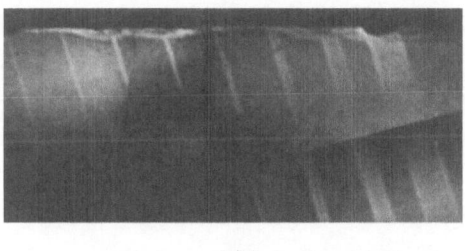

VIDEO — XRT

(b)

Fig. 6.7a and b. Reflection topogram of boron diffusion through oxide stripes of varying width; (a) optical micrograph (b) video X-ray topogram after *Rozgonyi* et al.

argument does not hold if Mo or Cu radiation is used. The lower absorption of the silicon target due to the shorter wavelength should decrease the resolution.

Chikawa and *Fujimoto* were the first investigators who, in 1968, employed a single-stage imaging method for viewing and recording of X-ray topographic images with sufficient resolution. They imaged individual dislocations in BeO crystals with a PbO vidicon type camera tube in their first investigation [6.27]. *Chikawa* and his colleagues have employed and improved their system continuously since that time. They reported direct imaging of stationary and moving dislocations in silicon crystals [6.40, 41] and Berg-Barrett imaging of IC patterns and slip bands in silicon crystal wafers [6.42]. The double-crystal technique was used to visualize dislocations, lattice parameter variations, and growth striations in silicon wafers [6.42]. Impressive observations were made of defects associated with growth of silicon crystals from the melt [6.43] and with dislocations generated by thermal stresses in silicon crystals [6.44].

Imaging System

Chikawa et al. preferred the Lang geometry. Their X-ray generator has an apparent focus size of 0.5 mm in the vertical direction and 1 mm in the horizontal direction. A geometrical resolution of the experimental setup of about 10 µm is sufficient for the limited resolution of the TV system (20–25 µm, as described later). Therefore, the distance between the X-ray source and the specimen was chosen as 1200 mm, since the distance from the specimen to the vidicon is

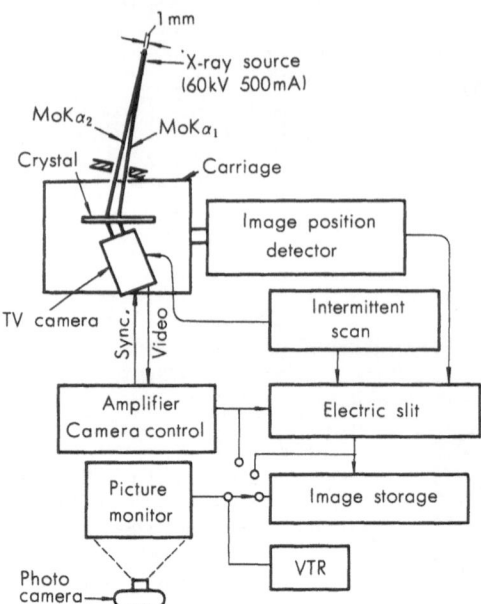

Fig. 6.8. Block diagram of the direct-display system developed by *Chikawa* et al.

required to be 30 mm for dynamic experiments with a furnace in *Chikawa*'s arrangement. The geometrical resolution in the vertical direction was then about 10 μm and in the horizontal direction about 3 μm by separating the K_{α_1} image from the K_{α_2} image. Diffracted intensities sufficient for direct viewing of defects require a high power rotating anode X-ray generator. *Chikawa* et al. used a Rigaku Denki RU 500 generator at 60 kV and 500 mA with a focus size of 0.5 mm × 10 mm. The principles of rotating anode generators are discussed in detail in Chapter 2 of this book.

A block diagram of the imaging system is shown in Fig. 6.8. The TV system consists of a TV camera unit with an X-ray sensitive vidicon-type camera tube, video amplifier, camera control unit, and TV monitor operating as a normal closed-circuit TV system. A standard 525-line, interlaced, 30-frame/s scanning system is employed. An intermittent scan which permits a charge storage on the photoconductive target before readout by the electron beam is applied for weak images. Two images due to the diffracted K_{α_1} and K_{α_2} beams, each with a width of 1 mm, are received by the TV camera as shown schematically in Fig. 6.8 after the adjustment of the crystal. These two band-shaped regions of the crystal are imaged simultaneously with a magnification of about 30 times on the TV monitor. *Chikawa* refers to these images as "direct view images". Larger areas of the crystal are imaged by selecting of the video signal due to the K_{α_1} image by the "electric slit" unit. This information is collected in the image storage unit while the carriage is moved a time of 3–10 s. The position of the "electric slit" is moved in synchronization with the carriage motion. The initial position, slit width, and traversing speed are adjusted while viewing the topographic images on the TV

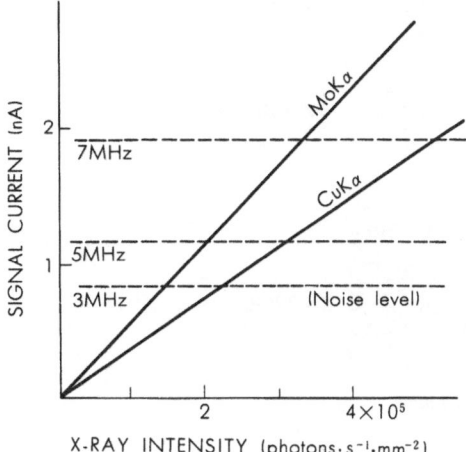

Fig. 6.9. Signal current of the PbO camera tube due to the incident X-ray intensity and noise level of the video amplifier (after *Chikawa* et al.)

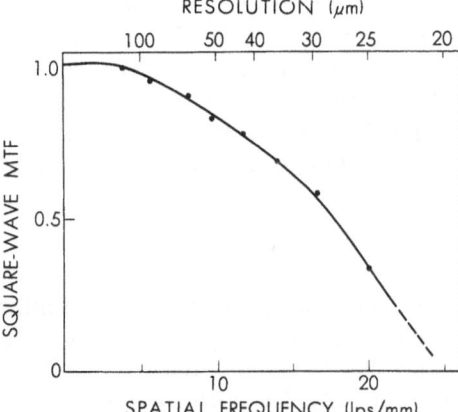

Fig. 6.10. Resolution of the PbO camera tube measured by the square-wave modulation transfer function (MTF) defined by

$$MTF = \frac{I_{omax} - I_{omin}}{I_{omax} + I_{omin}} \bigg/ \frac{I_{imax} - I_{imin}}{I_{imax} + I_{imin}},$$

where I_{imax} and I_{imin} are the maximum and minimum intensities of the incident image, and I_{omax} and I_{omin} are those of the output (after *Chikawa* et al.)

monitor. A Lang topogram of dimensions 9 mm × 13 mm which is obtained after each scan is displayed on the monitor. *Chikawa* refers to these images as "synthesized". Note that the pictures of crystal areas due to K_{α_1} and K_{α_2} wavelengths are imaged in the "direct image" while the "synthesized image" is made using the K_{α_1} wavelength only.

The construction and operation of the PbO camera tube are similar to a light sensing vidicon except for an X-ray sensitive target layer and a beryllium window. The camera tube was developed by *Nishida* and *Okamoto* who reported the characteristics in detail [6.45]. The sensitivity of the camera tube increases with the target voltage. The target voltage is limited to 50 V because higher voltages seem to reduce the tube life. The PbO layer should be thick for high sensitivity. However, the spatial resolution decreases with increasing thickness. Therefore, the thickness of the PbO layer was made about 15 μm to obtain good

ROTATING TARGET   SPECIMEN FURNACE

TV–CAMERA

CARRIAGE

Fig. 6.11. Experimental setup of the system of *Chikawa* et al.

sensitivity in combination with a good resolution. The efficiencies are 60 % and 78 % for Mo K_α (17 keV) and Cu K_α (8 keV) radiation, respectively. Figure 6.9 shows the sensitivity measured for Mo K_α and Cu K_α radiations. *Chikawa* et al. need more than 2×10^5 X-ray photons s^{-1} mm^{-2} in the diffracted beam to obtain a video signal higher than the noise level of the TV system used. The resolution capabilities of the camera tube are shown in Fig. 6.10 by a square wave modulation transfer function (MTF) measured with an amplifier having a bandwidth of 7 MHz. The limiting spatial resolution (at MTF = 5 %) is 20 to 25 μm.

Experiments

Figure 6.11 shows the experimental setup in Lang geometry. Topographic images of dislocations in a silicon wafer are reproduced in Fig. 6.12. Figure 6.12a shows a direct view image with the two band-shaped pictures due to K_{α_1} and K_{α_2} radiations. The black lines are dislocation contrasts. The horizontal band is the image of the electric slit. The wafer is adjusted so that the slit and the K_α images were parallel. The electric slit is then positioned on the K_{α_1} image as seen in Fig. 6.12b. A synthesized image is obtained by scanning the carriage (Fig. 6.12c). The video waveform along the dotted line in Fig. 6.12c is photographed in Fig. 6.12e. The peaks due to dislocation contrasts are visible. The intensity of the perfect region was about 4×10^5 Mo K_{α_1} photons s^{-1} mm^{-2}. Figure 6.12d was synthesized by using the intermittent scan mode with a time interval of 2.5/30 s. The video signal (Fig. 6.12f) becomes stronger due to the storage of TV frames by the intermittent scan. The same carriage velocity of 4 mm s^{-1} was applied to obtain the topograms in Figs. 6.12c and d. The image in Fig. 6.12d has a better image quality due to intermittent scanning. Figure 6.12g was synthesized with a carriage velocity of 1 mm s^{-1} and shows a further improvement in the image quality by increasing the integration time of the X-ray images. A comparison between the topograms in Figs. 6.12g and 6.12h which was recorded photographically gives a good correspondence.

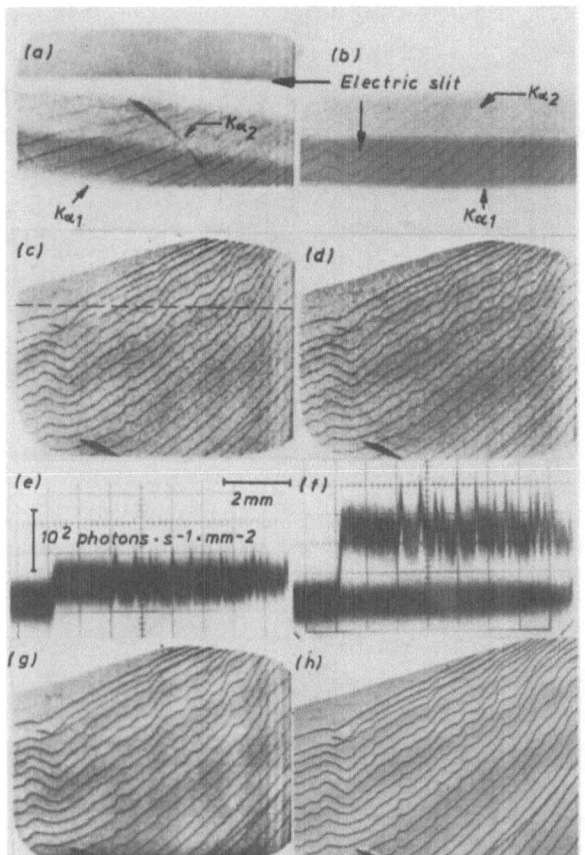

Fig. 6.12a–h. Video display of transmission topograms of a Si wafer (after *Chikawa* et al.). (a) Direct view image as seen instantaneously on the TV monitor; (b) superposition of the electrical slit; (c) synthesized image obtained in 3 s; (d) synthesized image obtained with the intermittent scan; (e) output signal of the video amplifier along dashed line in (c); (f) increased output signal due to intermittent scan; (g) synthesized image obtained in 10 s (standard scan); (h) X-ray topogram recorded on nuclear plate

A silicon wafer with IC devices was observed by Cu K_α radiation with a (422) asymmetric reflection in the Berg-Barrett method. A synthesized image was achieved with a carriage velocity of 1 mm s^{-1} (Fig. 6.13a). Figure 6.13b is a magnification of a pattern element in the central region of Fig. 6.13a. Figure 6.13c is a topogram of the same IC element recorded by a photographic plate. Two strips indicated by arrows are seen in Fig. 6.13c. They have a width and a separation of about 15 μm. These stripes are also seen in Fig. 6.13b. Therefore, *Chikawa* concluded that the TV tube should have a resolution of 15 μm in its central part for the experimental conditions used.

A silicon crystal was examined with Cu K_α radiation and (440) reflection in a double-crystal arrangement as shown in Fig. 6.14c. The width of the diffracted beam was broadened to 3 mm due to the asymmetrical reflection of the reference crystal used. The signal intensity received from the perfect regions was 1.5×10^5 Cu K_{α_1} photons s^{-1} mm^{-2}. Therefore, the video signal level was lower than the noise level, and it was impossible to see the direct image. *Chikawa* obtained a

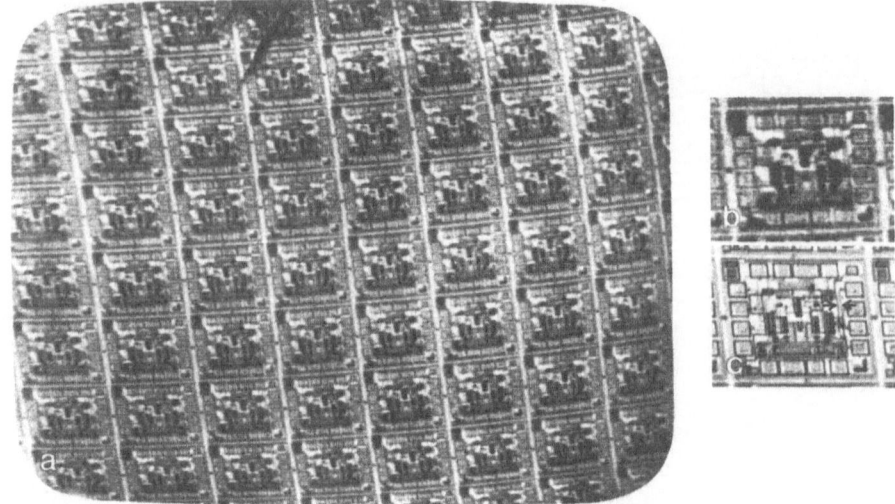

Fig. 6.13a–c. Video topograms of an IC wafer by the Berg-Barrett method after *Chikawa* et al. (a) synthesized image; (b) enlargement of an IC pattern in (a); (c) optical micrograph of the same pattern as (b). Arrows indicate two trips with 15 μm width and separation

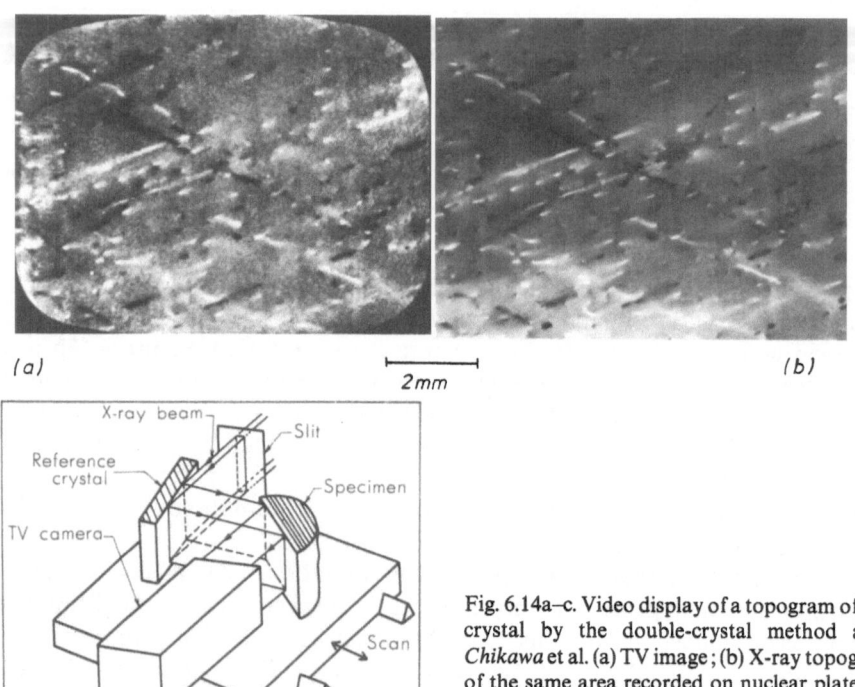

Fig. 6.14a–c. Video display of a topogram of a Si crystal by the double-crystal method after *Chikawa* et al. (a) TV image; (b) X-ray topogram of the same area recorded on nuclear plate; (c) schematic illustration of the double-crystal method

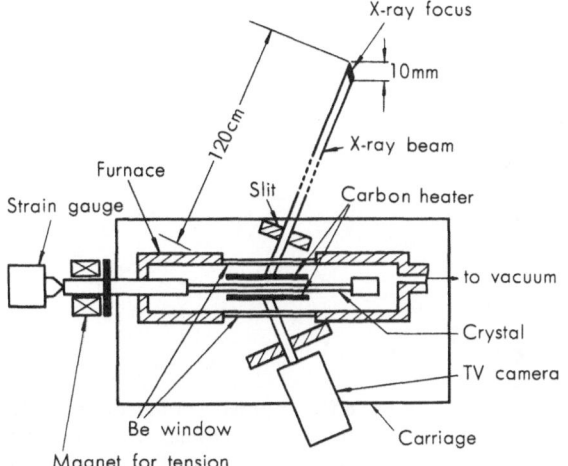

Fig. 6.15. Schematic drawing of the topographical method for observation of moving dislocations (after *Chikawa* et al.)

synthesized image of good quality (Fig. 6.14a) with an intermittent scan of a time interval of 0.5 s. The video image in Fig. 6.14a has no large loss of spatial resolution compared with the topogram in Fig. 6.14b which was recorded on a photographic plate.

Chikawa placed a 300 µm thick silicon wafer between two carbon-plate heaters in a vacuum furnace for observation of moving dislocations. The furnace was mounted on the carriage of the topographic camera in Lang arrangement as shown schematically in Fig. 6.15. The observation was achieved by the TV camera through the beryllium windows of the furnace. The temperature of the specimen was increased under an applied constant tensile stress. The dislocation motion at 950–1000 °C was recorded with Mo K_{α_1} radiation by a video tape recorder (VTR). The reproduced pictures from the VTR were photographed by a cine camera. Three successive TV pictures were superimposed on one movie frame. Figure 6.16 shows selected pictures of the movie and the sequence of the dislocation motion. The numbers at the top indicate time intervals (in seconds) between the pictures. The changes of the dislocation configuration and the directions of the dislocation motion are shown schematically in Fig. 6.16k–m. The dislocation velocity was measured to be 0.3 mm s^{-1} from the position change of the dislocation marked B. The shear stress on the slip plane of the dislocations was calculated to about 250 g mm^{-2}[6.41].

Crystal growth experiments have been started [6.41] and are still in progress [6.46].

Chikawa et al. showed that their system can be applied in a wide experimental range. All components of the X-ray and the TV system and their influences on the spatial resolution and the SNR were investigated. The time resolution is lowered by image storage with analog signal processing to obtain a better image quality.

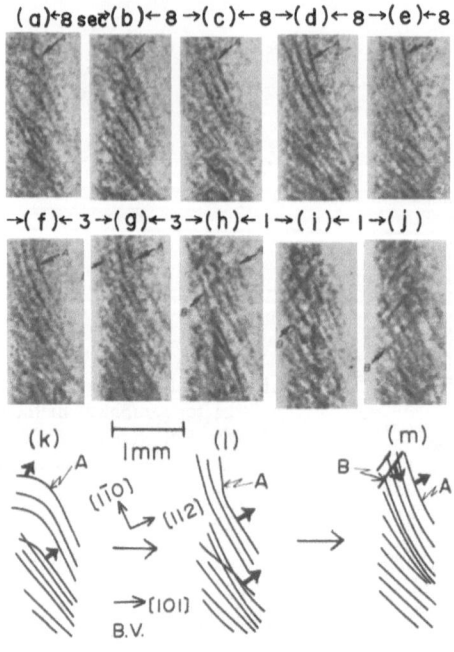

(a)←8 sec(b)←8→(c)←8→(d)←8→(e)←8

→(f)←3→(g)←3→(h)←1→(i)←1→(j)

Fig. 6.16. (a)–(j) selected frames from a 16-mm movie film showing video images of dislocation motion for 48 s. Numbers indicate time intervals, in seconds, between frames. (k)–(m) schematic drawings showing the motion sequence. The thick arrows indicate the directions of dislocation motion (after *Chikawa* et al.) *B.V.* indicates the Burgers vector

6.2.2 Multiple-Stage Imaging

Kohra and associates were the first investigators who used a multiple-stage imaging method with a good spatial resolution. They reported on experiments with a double-crystal arrangement to study long-range elastic strains in Sb-doped silicon crystals [6.47] and equal thickness fringes in Si-crystals [6.48]. The video system was also applied to diffraction topography of synthetic quartz and silicon web crystals [6.49, 50]. The high resolution arrangement which will be discussed in detail now was reported in 1972 [6.51].

Imaging System

The geometry of the experimental setup based on the divergent Laue method is shown in Fig. 6.17. White X-rays originating from a point focus source of high brilliancy enter the specimen through a small circular aperture. The specimen is positioned 7–13 mm from the X-ray source. X-rays which meet the specimen at the Bragg angles are diffracted and produce a number of Laue spots. Each of the Laue spots focuses at some distance from the specimen and then diverges. The Laue spots which carry topographic information of the specimen produce magnified diffraction images of the irradiated part of the specimen. The surface of the detector system is placed behind the focusing position to look at the magnified Laue spots. A beam stopper protects the detector surface from the

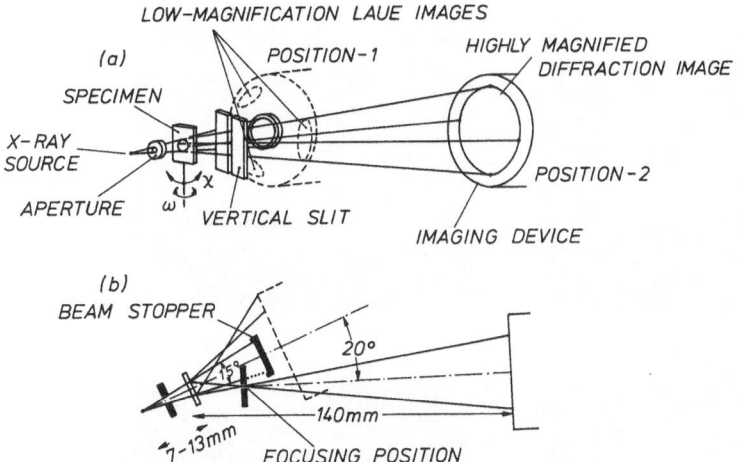

Fig. 6.17a and b. Schematic diagram of the experimental arrangement used by *Kohra* et al. (a) perspective view; (b) plan view

intense primary beam. Several low magnification Laue images are simultaneously observed by setting the detector system at position 1 in Fig. 6.17. The crystallographic orientation of the specimen should be readily known. The indices of the low magnification Laue spots are determined by their relative positions. One of the Laue spots is viewed in a higher magnification if the detector system is placed at position 2 in Fig. 6.17. A narrow vertical slit positioned at the focusing point reduces the scattered background radiation.

The topographic images thus observed undergo different magnifications in the horizontal and vertical directions. The resulting distortion of the image is insignificant if the distance from the X-ray source to the detector is large compared to that from the X-ray source to the specimen. The front of the detector system is placed about 140 mm from the specimen to study highly magnified Laue spots. The magnification is about 12 to 21 times at this position. The observed field of view is about 2 mm in diameter limited by the diameter of the incident aperture.

A Microflex X-ray source of Rigaku Denki Co. was used which operated at 50 kV and 230 μA with a tungsten target. The apparent focus size was 2.5 μm in the horizontal and 10 μm in the vertical direction. The X-ray wavelength used ranged from 0.28 Å to 1.5 Å. Wavelength ranges which contain characteristic radiation were usually avoided because they produce very strong lines in the observed topograms (see Fig. 6.18d).

A fluorescent screen converted X-rays into visible light in the first stage of the imaging system. The screen used had a thickness of about 50 μm and consisted of a mixture of ZnS and (Zn, Cd)S material activated by Ag. The screen was fiberoptically coupled to the photocathode of the TV camera tube. The camera tube was an image orthicon tube Shibaden HS 191. This camera tube has a

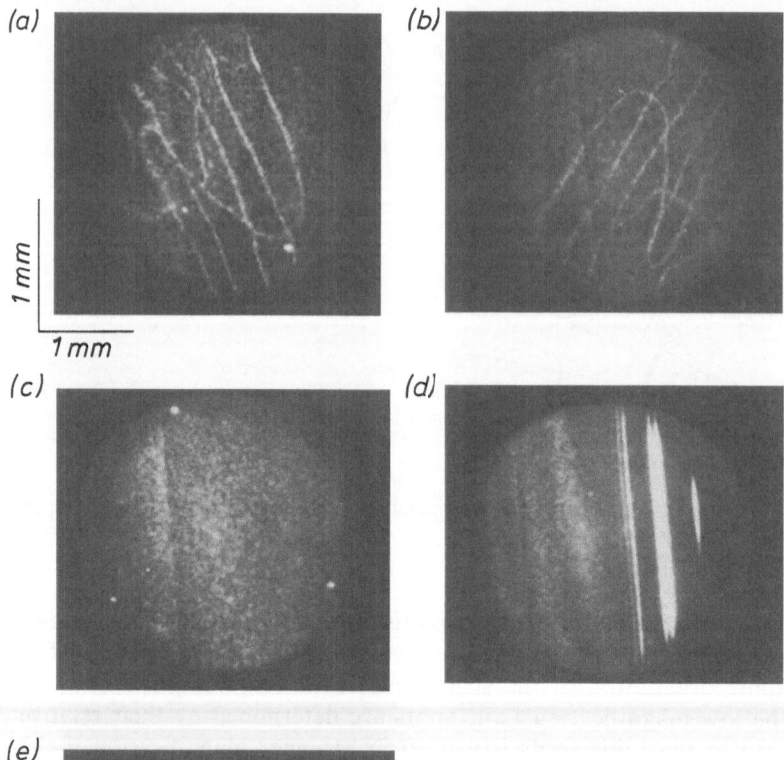

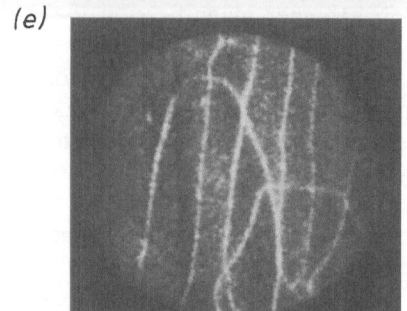

Fig. 6.18a–e. Video display of individual dislocation contrasts in a Si crystal obtained from various reflecting planes. (a) 3$\bar{1}\bar{1}$; (b) $\bar{1}3\bar{1}$; (c) $\bar{1}\bar{1}3$; (d) 11$\bar{1}$; (e) 2$\bar{2}$0 reflections. The strong vertical lines in (d) are caused by the tungsten characteristic lines of the *L*-series (after *Kozaki* et al.)

multialkali photocathode (S-20) with an active diameter of 42 mm and a MgO target. The imaging system possessed a sensitivity such that 3×10^4 8 keV X-ray photons s^{-1} mm^{-2} produced a video signal with 10:1 SNR. The target of the image orthicon camera allowed charge storage, and, therefore, the images could be electronically integrated. Integration for 1 s improved the sensitivity by more then a factor of 10. The spatial resolution of the system was limited to 300 μm for 70% of the MTF by the resolution of the fluorescent screen. This value corresponded to about 90 μm limiting resolution. The TV system used a 525-line scan with 30 frames per second and had a bandwidth of 7 MHz.

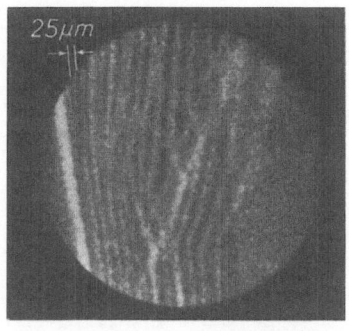

Fig. 6.19. Dislocation array in a Si web crystal; video topogram in ($2\bar{2}0$) reflection (after *Kozaki* et al.)

500 μm

Experiments

The first experiment concerned a silicon wafer of 0.2 mm thickness. The incident X-ray beam was approximately normal to the (111) surface of the wafer. Several Laue images were observed in a higher magnification as shown in Fig. 6.18. These pictures were reproduced by photographing the TV monitor with an exposure time of 5 s. A TV display of dislocation arrays is imaged in Fig. 6.19. This picture was observed in the ($2\bar{2}0$) reflection of a silicon web crystal 98 μm thick. Dislocations as close as 25 μm can be seen distinctly resolved. An IC pattern deposited on a silicon wafer 0.2 mm thick was observed in a third experiment [6.51].

More enlargement of the Laue spots to increase the spatial resolutions is impossible because the diffracted X-ray beam becomes too weak to produce a sufficient video signal. The method of *Kozaki* et al. cannot be employed to observe reflection type topograms because polychromatic X-rays diffracted from the specimen in the reflection geometry focus at imaginary positions only and scattered background radiation can no longer be suppressed by a narrow slit at the focusing position.

The present author and his colleagues reported in 1975 about a high resolution direct-display system which works with a multiple-stage imaging method [6.52].

Imaging System

Figure 6.20 shows a diagram of the system. An X-ray beam enters the sample at a distance of about 150 mm from the X-ray source. The horizontal and the vertical divergences of the primary beam are limited by a double slit. The specimen is set in Bragg condition for the desired (hkl) reflection. A diaphragm behind the specimen stops the primary beam. The diffracted intensity is converted into visible light by a fluorescent screen. This screen is positioned a distance of about 3 mm from the specimen. The geometry used has unusually short distances compared with the standard Lang method for transmission topography. The K_{α_1} radiation is diffracted from another portion of the crystal than the K_{α_2} radiation

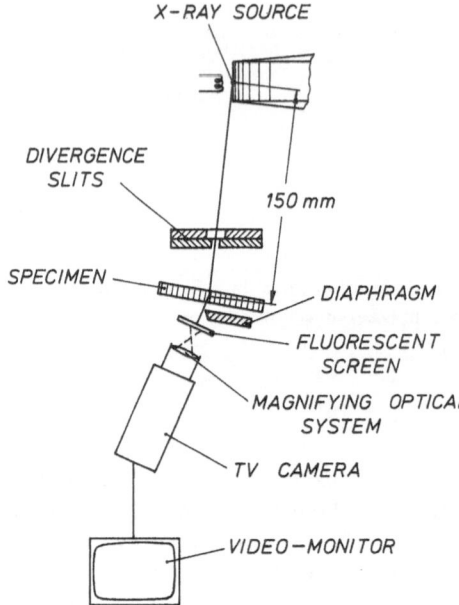

Fig. 6.20. Schematic diagram of the video display X-ray topography system (after *Hartmann* et al.)

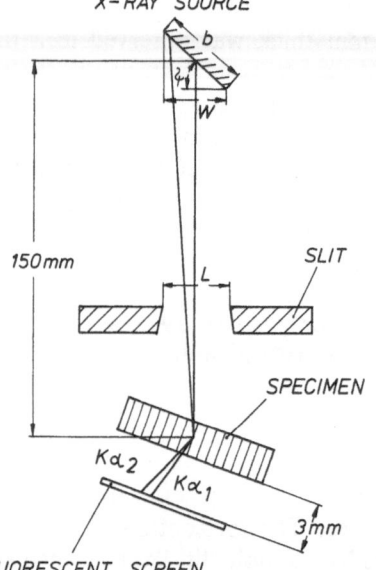

Fig. 6.21. Schematic drawing of the geometrical imaging due to K_{α_1} and K_{α_2} radiation

in the regular transmission arrangement due to the large distances from X-ray source to the specimen. Therefore, the Lang method allows a separation of the images from crystal regions which are imaged from the K_{α_1} and K_{α_2} radiation, respectively, and a suppression of the K_{α_2} image by a narrower incident slit before the specimen. The crystal region which is irradiated by the X-ray beam diffracts K_{α_1} and K_{α_2} radiation at the same time due to the small distances in our method. Therefore, we obtained a broadening of defect contrasts due to the different Bragg angles of the K_{α_1} and K_{α_2} radiation and a superposition of the corresponding images. This broadening effect is drawn schematically in Fig. 6.21 and should be discussed in more detail because it is important for the horizontal resolution.

The Bragg condition $\lambda = 2d_{hkl} \sin \Theta_B$ after differentiation gives a difference in the Bragg angles caused by the different wavelengths expressed by

$$\Delta\lambda = 2d_{hkl} \cos \Theta_B \Delta\Theta_{1-2} \tag{6.2}$$

or

$$\Delta\Theta_{1-2} = \frac{\Delta\lambda}{2d_{hkl}(1 - \sin^2 \Theta_B)^{1/2}} . \tag{6.3}$$

One obtains from the Bragg equation $\sin^2 \Theta_B = (\lambda/2d_{hkl})^2$. Therefore, the difference $\Delta\Theta$ can be written

$$\Delta\Theta_{1-2} = \frac{\Delta\lambda}{2d_{hkl}\left[1 - \left(\dfrac{\lambda}{2d_{hkl}}\right)^2\right]^{1/2}} . \tag{6.4}$$

The broadening v at a distance M from the specimen is then given by

$$v = M\Delta\Theta_{1-2} . \tag{6.5}$$

The distance between the (220) lattice planes d_{220} is 1.97 Å in a silicon crystal and the wavelength spread $\Delta\lambda$ of Mo K_α radiation is 0.00428 Å. We obtain $\Delta\Theta_{1-2} = 1.103 \times 10^{-3}$ rad with these values. A broadening v of defect contrasts of about 3.3 µm is achieved with a distance $M = 3$ mm. The basic geometrical resolution of the system can be calculated by the expression [6.53]

$$D_h = \frac{M\Delta\Theta}{\sin(\alpha + \Theta_B)} \tag{6.6}$$

if we assume that the crystal is imaged by the K_α radiation only. The distance between specimen and fluorescent screen is denoted M. The geometrical divergency of the K_{α_1} radiation which can be diffracted from the crystal is named $\Delta\Theta$, the Bragg angle for a (hkl) reflection is Θ_B, and α is the angle between the normal of the reflecting plane and the normal of the specimen surface. The divergence of the K_{α_1} radiation is calculated to be about 2 mrad. The distance M is 3 mm. We obtain $\Theta_B = 10.64$ deg and $\alpha = 90$ deg for Mo K_{α_1} radiation and (220) reflection in a silicon wafer with (111) surface. The basic geometrical resolution in horizontal direction is then about 6.1 µm. The basic resolution and the broadening should be added to achieve the overall resolution of the system. The system has then a resolution of about 9–10 µm in the horizontal direction.

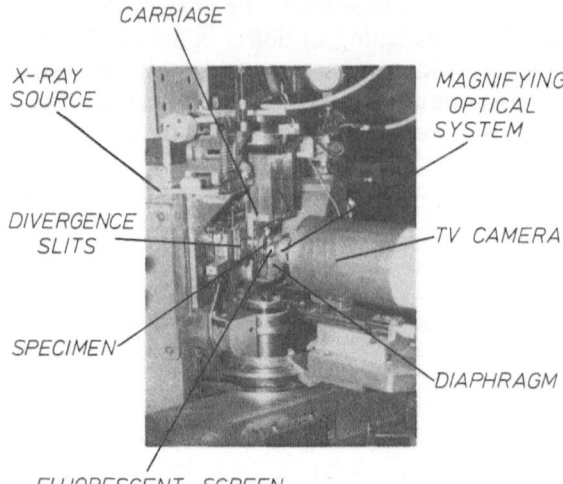

CARRIAGE

X-RAY
SOURCE

MAGNIFYING
OPTICAL
SYSTEM

DIVERGENCE
SLITS

TV CAMERA

SPECIMEN

DIAPHRAGM

Fig. 6.22. Experimental setup of
the system of *Hartmann* et al.

FLUORESCENT SCREEN

The geometrical resolution in the vertical direction is determined in the same way as in the Lang method by the expression $D_v = MH/L$ (6.1). The resolution in the vertical direction is about $3\,\mu m$ with $M = 3\,mm$, $L = 150\,mm$ and a vertical focus size of $0.15\,mm$. We used the rotating anode X-ray generator RU-500 from Rigaku Denki Co. for our experiments. The power of the generator was $60\,kV$ and $80\,mA$ with a fine focus size of $0.1 \times 2\,mm^2$. As shown in Fig. 6.20, the detector system is composed of a fluorescent screen, a magnifying optical system, the TV camera, and the TV monitor. The screen employed had a thickness of about $10\text{–}15\,\mu m$. It consisted of Zn_2SiO_4 material with a grain size of about $1\text{–}2\,\mu m$. The optical system magnified the visible pattern on the fluorescent screen into the target of a low-light-level TV camera. The optical magnification was about 6 times. The goal of a resolution of $10\,\mu m$ for the detector system made this optical magnification indispensable because the TV camera tube had a limiting resolution of about $25\text{–}30\,\mu m$. The electronic magnification between the target of the TV camera and the TV monitor was about 15 times. Therefore, we saw an image of a crystal area of about $0.35 \times 1.5\,mm^2$ in an overall magnification of 90 times.

Experiments

Figure 6.22 shows the real experimental setup. A transmission topogram is imaged in Fig. 6.23. This topogram was taken with the standard Lang technique and Mo K_{α_1} radiation of a $0.5\,mm$ thick silicon wafer in (220) reflection. Figure 6.24 shows photographically magnified portions of the topogram. These portions are encircled in Fig. 6.23. The three oblique contrasts in Fig. 6.24a have a distance of about $9\,\mu m$ and $19\,\mu m$, respectively. The dislocation loop in Fig. 6.24b has a maximum diameter of about $17\,\mu m$ and Fig. 6.24c shows another example of dislocation contrasts. The same contrasts as in Fig. 6.24 are imaged in

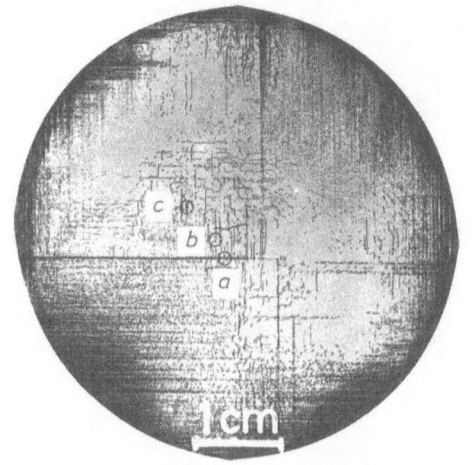

Fig. 6.23. Standard topogram of a 0.5 thick Si wafer recorded with the Lang method in (220) reflection with MoK_{α_1}; encircled portions indicate the magnified pictures in Fig. 6.24

0.35mm   0.35mm
  (a)   (b)   0.35mm
                         (c)

Fig. 6.24a–c. Photographically magnified portions of Fig. 6.23 (after *Hartmann* et al.). (a) The three oblique contrasts have a distance of about 9 μm and 19 μm, respectively. (b) The loop has a maximum diameter of about 17 μm; (c) another example of dislocation contrasts

Fig. 6.25 but photographed directly from the video monitor with an exposure time of 0.5 s. Fine details of individual dislocations are evident in this reproduction. We concluded that dislocation contrasts with a separation of about 10 μm can be resolved in our comparison with the topogram taken by the Lang method. Dynamical experiments are still in progress. An investigation of the behavior of ion implanted silicon wafers with increasing temperatures up to 1200 °C will be prepared.

The system described above can be employed in transmission and reflection experiments. Disadvantages are the small field of view and the low absorption coefficient of the fluorescent screen. The high spatial resolution of a few μm and the possibility to vary the optical magnification over a wide range are the advantages.

6.3 Summary

The properties of the described systems are discussed in this section. The advantages and disadvantages in relation to good spatial and time resolution and the signal-to-noise ratio are pointed out.

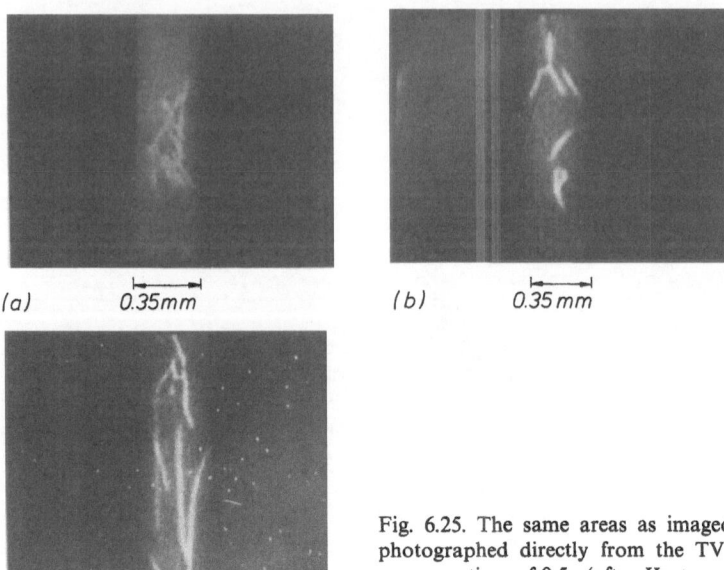

(a)   0.35mm

(b)   0.35mm

(c)   0.35mm

Fig. 6.25. The same areas as imaged in Fig. 6.24 but photographed directly from the TV monitor with an exposure time of 0.5 s (after *Hartmann* et al.)

The signal-to-noise ratio (SNR) should be not smaller than 2 to recognize the defect image in the perfect region. Following *Chikawa* [6.41] the SNR is given by the expression

$$R = \varepsilon C (\eta_0 v_p t)^{1/2} \geqq k \qquad (6.7)$$

where ε^2 is a square-shaped picture element determined by the resolution of the system. The perfect region of a crystal diffracts v_p X-ray photons s^{-1} mm^{-2} and η_0 is the absorption coefficient of the X-ray sensitive layer of the system. The integration time of the pictures is named t, and k is the threshold SNR of about 1 to 5. The contrast of the defect image is defined by $C = (S_d - S_p)/S_p$. The signal level of the defect image is expressed by S_d, and S_p defines the mean signal level obtained from each square-shaped picture element $\varepsilon \times \varepsilon$ mm^2 for t seconds from the perfect crystal. The mean signal level S_p can be calculated by

$$S_p = \eta_0 v_p \eta_1 \eta_2 \ldots \eta_s \varepsilon^2 t \qquad (6.8)$$

where η_1 to η_s are conversion, absorption, or amplification factors which depend on the system used. An absorbed X-ray photon produces a number $\eta_1 > 1$ of electrons or visible photons due to the large photon energy of X-rays; additionally there is an amplification factor η_s of the photon flux of an amplifier, for example. The defect signal level S_d can be calculated in a similar way.

It is obvious from (6.7) that a direct-display system should have a high diffracted intensity and a high absorption coefficient of the X-ray sensitive target to obtain a good spatial resolution ε and a good visibility of defect contrasts determined by the SNR. The spatial resolution can be increased by storage of pictures at the expense of time.

Rozgonyi et al. had a favorable absorption of about 40% of the $Cr K_\alpha$ radiation. The diffracted intensity was limited by the conventional X-ray generator. The contrasts of the metallized areas seem to be good when compared with dislocation contrasts in the Berg-Barrett method. The images were displayed every 1/5 s on the TV monitor with a resulting storage time of 0.2 s. The analog storage allowed a superposition of about 6 TV frames. These selected conditions gave a spatial resolution of about 20 μm. The resolution in time and space should rapidly decrease if the detector system is employed in transmission experiments with $Mo K_\alpha$ radiation.

Chikawa et al. made the most extensive investigations about image quality and noise characteristics of their system. They obtained a resolution ε of about 30 μm using a frame time $t = 1/30$ s. The absorption of the PbO layer η_0 was 0.6 with a measured SNR of about 2 from Fig. 6.12e. The measured resolution is in good agreement with the limiting resolution of about 25 μm of the PbO vidicon tube (see Fig. 6.10). The first dynamical X-ray topography experiments were made with this system. The advantages of the method are the large absorption of the PbO target for radiations which are mostly used in X-ray topography (60% and 78% for $Mo K_\alpha$ and $Cu K_\alpha$, respectively) and a large field of view of $2\,mm \times 13\,mm$ of the direct image. The disadvantage is that the resolution is limited to 25–30 μm.

Kozaki et al. used a special technique which can be employed in transmission experiments only. The method needs a microfocus X-ray generator with high luminosity and wavelength ranges without characteristic lines. The detector system had a high sensitivity such that 3×10^4 photons $s^{-1}\,mm^{-2}$ already gave a SNR of about 10. This sensitivity was obtained by a thick (50 μm) fluorescent screen. Therefore, the resolution was limited to 90 μm. The high thickness should be indispensable because the absorption of the wavelengths (0.28–1.5 Å) used by the fluorescent material $[ZnS + (Zn, Cd)S]$ is only about 5–10%. The resolution of 25 μm was produced by an X-ray optical magnification of about 21 times. A stronger magnification can be achieved by increasing the distance from specimen to the detector system. The system should obtain, therefore, a higher resolution. The X-ray beam received by the detector system becomes too weak if the magnification is too high. The video signal lies then under the noise level of the video system. The resolution of the method could be increased by a fluorescent material with a higher resolution and efficiency in the short wavelength range.

The investigation of the noise characteristics and other properties of the system which was developed by the present author and co-workers has not been finished at this time. The geometrical arrangement can be varied and adapted in a wide range to fulfill the special conditions of specific experiments in transmission and reflection geometry. Electronic image storage and evaluation are in preparation. Also, dynamical experiments are planned. Disadvantages are the small field of view which is limited to $0.4\,mm \times 2\,mm$ and the small distance between specimen and fluorescent screen. The advantages are the high resolution of the fluorescent material without loss of efficiency and the possible magnification of the optical pattern on the fluorescent screen to overcome the limited resolution of the TV system (25–30 μm). The overall magnification can be varied from 10 to 200 times in order to see fine details of individual dislocations. A fluorescent material with a better sensitivity but the same resolution should be developed to obtain better image quality and to visualize weak contrasts at a higher magnification.

Note that none of the reported systems imaged dislocation contrasts in the Berg-Barrett arrangement. Only topograms of IC patterns are imaged with this technique. This fact may be caused by inherently low SNR of the Berg-Barrett technique.

6.4 Future Trends

6.4.1 X-Ray Sources

A higher X-ray intensity results in an increased SNR in the video signal as obviously seen from (6.7) and, therefore, a better visibility of defect contrasts can be achieved. Rotating anode X-ray generators which have been used in the systems described had a total power of 60 kV and 500 mA at a focus size of 0.5 mm × 10 mm. A new high-brilliancy rotating anode machine is available now from Rigaku Denki Co. The total power is 60 kV and 0.06 mA at a focus size of 0.1 mm × 1 mm. This new X-ray generator has, therefore, an increased intensity factor of about 2 compared with the generators used until now. Intense radiation originating from synchrotrons or, better, storage rings has an X-ray photon flux more than a factor 10 higher than the X-ray rotating anode generators [6.54]. Another advantage of synchrotron radiation is horizontal and vertical divergence smaller than 1 mrad on the entire X-ray beam area in the wavelength range of interest below 3 Å. Therefore, the full width of the X-ray beam can be used to image crystal regions, and the field of view becomes larger than that obtained with conventional X-ray machines.

6.4.2 Detector Systems

The detector system should have a high absorption and a spatial resolution of a few μm simultaneously with a time resolution better than 1 s [see (6.7)] to achieve good image quality and to observe the dynamical behavior of defects. The PbO tube in *Chikawa*'s system has an absorption of 60% in the wavelength range of interest but the spatial resolution is limited to 25–30 μm. The technology seems to be limited and the spatial resolution cannot be exceeded. Another interesting development is the channel plate [6.55]. The resolution has been limited to 30 μm by the channel pitch. *Kellogg* et al. [6.56] reported a high resolution X-ray detector with a spatial resolution of 10 μm. The X-ray detector consists of microchannel plates as a combined photocathode surface and imaging photoelectron multiplier, and a crossed wire grid as a two-dimensional position sensitive detector. The disadvantage is that the detector system has sufficient quantum efficiencies up to 29 % in the soft X-ray range (0.28–3 keV) only. The microchannel plates can be important if the further development achieves better absorption and resolution for the short wavelengths.

The most important part of a multiple-stage imaging detector is the incident fluorescent screen. The screen material should have a high absorption of X-ray photons and a good conversion efficiency. The spatial resolution depends on the grain size of the fluorescent material. The grain size should not exceed a few μm without loss of efficiency. The Zn_2SiO_4 material used by the present author has no loss of efficiency down to a grain size of 0.5 μm, but the absorption of X-ray photons is only about 5 %. Further developments are still in progress in our laboratory in order to increase the absorption of X-ray photons of the fluorescent material with retention of a resolution of a few μm.

6.4.3 Electronic Image Processing

All methods described provide the possibility of storing TV frames to obtain better image quality. This storage is achieved by analog methods in the reported systems. Digital image processing should be increasingly employed in the future because the digital method has several advantages. The storage time can be varied over a large range with exact image correlation. Therefore, time resolution can be traded off against spatial resolution. This trade-off can be very important because many dynamical processes of defect configuration pass slowly, but it is necessary to have a high spatial resolution to detect fine details of defects during the dynamical processes. Another advantage of digital image processing is the exact covering of successive images allowing image subtraction. Variations of the defect structure are imaged only by image subtraction. Live topography with electronic image processing could become a powerful tool for quantitative and qualitative observation of the dynamical behaviour of lattice defects.

Acknowledgements. The author would like to thank the various investigators whose systems he has discussed for supplying him with technical information and copies of original figures. He is grateful to H.-J. Queisser for critical discussions. Finally, he would like to thank Mrs. A. Vierhaus for typing and preparing the manuscript, and G. Markewitz for preparing the photographs.

References

6.1   W. Berg: Naturwiss. **19**, 391 (1931)
6.2   C. S. Barrett: Trans. AIME **161**, 15 (1945)
6.3   A. R. Lang: J. Appl. Phys. **29**, 597 (1958)
6.4   A. R. Lang: Acta Cryst. **12**, 249 (1959)
6.5   H. Barth, R. Hosemann: Z. Naturforsch. **13a**, 792 (1958)

6.6   J.B.Newkirk: Trans. Met. Soc. AIME **215**, 483 (1959)

6.7   S.B.Austerman, J.B.Newkirk: "Experimental Procedures in X-Ray Diffraction Topography", in *Advances in X-Ray Analysis*, ed. by J.B. Newkirk, G.R. Mallett, Vol. 10 (Plenum Press, New York 1967) pp. 134—152

6.8   K.Kohra, H.Hashizume, J.Yoshimura: Japan. J. Appl. Phys. **9**, 1029 (1970)

6.9   K.Kohra: "Multiple Crystal Arrangements". Paper presented at the International Summer School on X-Ray Dynamical Theory and Topography, Limoges, France, August 1975

6.10 A.R.Lang: Recent Applications of X-Ray Topography, in *Modern Diffraction and Imaging Techniques in Material Science*, ed. by S. Amelinckx, R. Gevers, G. Remaut, J. van Landuyt (North-Holland Publishing Company Amsterdam, London 1970) pp. 407—479

6.11 A.Authier: "Contrast of Dislocation Images in X-Ray Transmission Topography", in *Advances in X-Ray Analysis*, ed. by J.B. Newkirk, G.R. Mallett, Vol. 10 (Plenum Press, New York 1967) pp. 9—31

6.12 B.K.Tanner: *X-Ray Diffraction Topography*, 1st ed. (Pergamon Press Oxford, New York, Toronto, Sydney, Paris, Frankfurt 1976)

6.13 H.Schaumburg: Phil. Mag. **25**, 1429 (1972)

6.14 A.George, C.Escaravage, G.Champier, W.Schröter: Phys. Stat. Sol. (b) **53**, 483 (1972)

6.15 A.George, C.Escaravage, W.Schröter, G.Champier: Cryst. Latt. Defects **4**, 29 (1973)

6.16 H.E.Krüger: *Hochtemperatur-Röntgentopographie an Silizium und Galliumarsenid*, Thesis (Rheinisch-Westfälische Technische Hochschule Aachen 1976)

6.17 V.G.Eremenko, V.I.Nikitenko: Phys. Stat. Sol. (a) **14**, 317 (1972)

6.18 H.Weyerer: Acta Cryst. **14**, 771 (1961)

6.19 G.W.Goetze, A.Taylor: "Some Recent Developments in the Direct Viewing and High-Speed Recording of X-Ray Diffraction Patterns", in *Advances in X-Ray Analysis*, ed. by W.M. Mueller, Vol. 5 (Plenum Press New York 1962) pp. 86—93

6.20 G.W.Goetze, A.Taylor: Rev. Sci. Instr. **33**, 353 (1962)

6.21 R.Carlson, T.C.Furnas Jr., D.W.Beard: Acta Cryst. **16**, A148 (1963) (supplement)

6.22 S.W.Kennedy: Nature **210**, 936 (1966)

6.23 T.D.Mokul'skaya, M.A.Mokul'skii: Sov. Phys. Doklady **11**, 132 (1966)

6.24 U.W.Arndt, B.K.Ambrose: IEEE Trans. NS-15, 92 (1968)

6.25 R.E.Green Jr., K.Reifsnider: J. Metals **20**, 83A (1968)

6.26 K.Reifsnider, R.E.Green Jr.: Rev. Sci. Instr. **39**, 1651 (1968)

6.27 J.Chikawa, I.Fujimoto: Appl. Phys. Letters **13**, 387 (1968)

6.28 A.N.Chester, T.C.Loomis, M.M.Weiss: Bell System Tech. J. **48**, 345 (1969)

6.29 A.N.Chester, F.B.Koch: "Instantaneous Display of X-Ray Diffraction Using a Diode Array Camera Tube", in *Advances in X-Ray Analysis*, ed. by C.S. Barrett, J.B. Newkirk, G.R. Mallett, Vol. 12 (Plenum Press, New York 1969) pp. 165—173

6.30 E.S.Meieran, J.K.Landre, S.O'Hara: Appl. Phys. Letters **14**, 368 (1969)

6.31 A.R.Lang, K.Reifsnider: Appl. Phys. Letters **15**, 258 (1969)

6.32 M.Blamoutier: "Un Tube de Prise de Vues Sensible aux Rayons X", in *Advances in Electronics and Electron Physics*, ed. by L. Marton, Vol. 28A (Academic Press New York 1969) pp. 273—280

6.33 B.Driard: "Contrôle des Monocristaux par Tube Intensificateur de Luminace", in *Advances in Electronics and Electron Physics*, ed. by L. Marton, Vol. 28B (Academic Press New York 1969) pp. 931—937

6.34 J.Ball, T.C.Furnas Jr.: Paper presented at the American Crystallographic Association Meeting, New Orleans 1970

6.35 R.E.Green Jr.: "Electro-Optical Systems for Dynamic Display of X-Ray Diffraction Images", in *Advances in X-Ray Analysis*, ed. by C.S. Barrett, J.B. Newkirk, C.O. Ruud, Vol. 14 (Plenum Press New York 1971) pp. 311—337

6.36 R.E.Green Jr.: "Direct Display of X-Ray Topographic Images", in *Advances in X-Ray Analysis*, ed. by H.F.McMurdie, C.S.Barrett, J.B.Newkirk, C.O.Ruud, Vol. 20 (Plenum Press New York 1977) pp. 221—235

6.37 W.Eckers, H.Oppolzer: Siemens Forsch.- u. Entwicklungs-Ber. **6**, 47 (1977)

6.38 G. A. Rozgonyi, S. E. Haszko, J. L. Statile: Appl. Phys. Letters **16**, 443 (1970)
6.39 G. A. Rozgonyi: private communication
6.40 J. Chikawa, I. Fujimoto, T. Abe: Appl. Phys. Letters **21**, 295 (1972)
6.41 J. Chikawa, I. Fujimoto: Nippon Hoso Kyokai (NHK) Technical Research Laboratories Technical Monograph No. 33, March 1974
6.42 J. Chikawa, I. Fujimoto, S. Endo, K. Mase: X-Ray Television Topography for Quick Inspection of Si Crystals, in *Semiconductor Silicon*, ed. by H. R. Huff, P. R. Burgess (Electrochemical Society Softbound Symposium Series Princeton N.J. 1973) pp. 448—458
6.43 J. Chikawa: J. Cryst. Growth **24/25**, 61 (1974)
6.44 J. Chikawa: Live Topography Using a Television System; Paper presented at the International Summer School on X-Ray Dynamical Theory and Topography, Limoges, France, August 1975
6.45 R. Nishida, S. Okamoto: Rep. Res. Inst. Electron., Shizuoka University **1**, 21, 185 (1966)
6.46 J. Chikawa: private communication
6.47 H. Hashizume, K. Kohra, T. Yamaguchi, K. Kinoshita: Appl. Phys. Letters **18**, 213 (1971)
6.48 H. Hashizume, H. Ishida, K. Kohra: Japan. J. Appl. Phys. **10**, 514 (1971)
6.49 K. Kohra: "Dynamical Asymmetric Diffraction and Its Applications to X-Ray Optical Systems", in *Proceedings of the Sixth International Conference on X-Ray Optics and Microanalysis*, ed. by G. Shinoda, K. Kohra, T. Ichinokawa (University of Tokyo Press, Tokyo 1972) pp. 35—45
6.50 H. Hashizume, K. Kohra, T. Yamaguchi, K. Kinoshita: "A Few Applications of an X-Ray Television System to Diffraction Studies", in *Proceedings of the Sixth International Conference on X-Ray Optics and Microanalysis*, ed. by G. Shinoda, K. Kohra, T. Ichinokawa (University of Tokyo Press, Tokyo 1972) pp. 695—701
6.51 S. Kozaki, H. Hashizume, K. Kohra: Japan. J. Appl. Phys. **11**, 1514 (1972)
6.52 W. Hartmann, G. Markewitz, U. Rettenmaier, H. J. Queisser: Appl. Phys. Letters **27**, 308 (1975)
6.53 M. Yoshimatsu: *X-Ray Diffraction Micrography—The Lang Method*, internal information of Rigaku-Denki Co., Tokyo
6.54 C. Kunz: "Perspectives of Synchrotron Radiation" (Report on a Panel Discussion), in *Vacuum Ultraviolet Radiation Physics-Proceedings of the Fourth International Conference on Vacuum Ultraviolet Radiation Physics*, ed. by E. E. Koch, R. Haensch, C. Kunz (Vieweg, Braunschweig 1974) pp. 753—771
6.55 M. Ando, S. Hosoya, K. Namikawa: J. Appl. Cryst. **9**, 269 (1976)
6.56 E. Kellogg, P. Henry, S. Murray, L. van Speybroeck, P. Bjorkholm: Rev. Sci. Instr. **47**, 282 (1976)

Additional References with Titles

Chapter 3

G. Boivin: Use of a Fresnel zone plate for optical image formation with short wavelength radiation. Appl. Opt. **16**, 1070 (1977)

A. Franks: X-ray Optics Science Progress, Oxf. **64**, 371 (1977)

E. Gipstein, A.C. Ouano, D.E. Johnson, O.U. Need III: Parameters affecting the sensitivity of Poly (methyl methacrylate) as a positive lithographic resist. Polymer Engineering and Science **17**, 396 (1977)

K. Murase, M. Kakuchi, S. Sugawara: Newly developed electron and X-ray resist materials. Intern. Conf. on Microlithography (Paris 1977) p. 261

Chapter 4

1. P.A. Bezirganyan: A three-wave method for the measurement of the length of a wave-train of X-ray radiation. Phys. stat. sol. (a) **40**, K 77 (1977)
2. U. Bonse, W. Graeff: On the possibility of measuring the length of X-ray or neutron wave trains with the three-beam case interferometer. Submitted to phys. stat. sol. (a) (1977)
3. D.C. Creagh: Determination of the mass attenuation coefficients and anomalous dispersion corrections for calcium for X-ray wavelengths from $IK\alpha_1$ to $CuK\alpha_1$. Phys. stat. sol. (a) **39**, 705 (1977)
4. H. Kaiser, H. Rauch, W. Bauspiess, U. Bonse: Measurement of the coherent neutron scattering length of ^{3}He by neutron interferometry. Submitted to Phys. Lett. B (1977)
5. H. Rauch, D. Petrascheck: "Dynamical Diffraction on Perfect Crystals and its Application in Neutron Physics", in *Neutron Diffraction*, ed. by H. Dachs, Topics in Current Physics; in preparation (Springer, Berlin, Heidelberg, New York 1978)
6. H. Rauch, E. Seidl, A. Zeilinger, W. Bauspiess, U. Bonse: Hydrogen detection in metals by neutron interferometry. Submitted to J. Appl. Phys. (1977)

Subject Index

Applied Physics

A monthly journal

Board of Editors	**S. Amelinckx,** Mol. · **V. P. Chebotayev,** Novosibirsk **R. Gomer,** Chicago, III. · **H. Ibach,** Jülich **V. S. Letokhov,** Moskau · **H. K. V. Lotsch,** Heidelberg **H. J. Queisser,** Stuttgart · **F. P. Schäfer,** Göttingen **A. Seeger,** Stuttgart · **K. Shimoda,** Tokyo **T. Tamir,** Brooklyn, N.Y. · **W. T. Welford,** London **H. P. J. Wijn,** Eindhoven
Coverage	application-oriented experimental and theoretical physics: *Solid-State Physics* *Quantum Electronics* *Surface Physics* *Laser Spectroscopy* *Chemisorption* *Photophysical Chemistry* *Microwave Acoustics* *Optical Physics* *Electrophysics* *Integrated Optics*
Special Features	**rapid** publication (3–4 months) **no** page charge for **concise** reports prepublication of titles and abstracts **microfiche** edition available as well
Languages	Mostly English
Articles	original reports, and short communications review and/or tutorial papers
Manuscripts	to Springer-Verlag (Attn. H. Lotsch), P.O. Box 105 280 D-69 Heidelberg 1, F.R. Germany Place North-American orders with: Springer-Verlag New York Inc., 175 Fifth Avenue, New York. N.Y. 10010, USA

Springer-Verlag
Berlin Heidelberg New York

Springer-Series in Solid-State Sciences

Editors: M. Cardona, P. Fulde, H. J. Queisser (Max-Planck-Institute for Solid-State Research, Stuttgart)

This book series is devoted to graduate-level monographs, textbooks, and (rarely) proceedings in the fields of solid-state physics, solid-state technology and solid-state chemistry. Semiconductor physics and technology as well as basic surface physics will also be considered.

Springer-Verlag
Berlin
Heidelberg
New York

Volume 1
C. P. Slichter, University of Illinois, Urbana, IL, USA

Principles of Magnetic Resonance

2nd, revised and expanded edition. 112 figures. Approx. 360 pages. 1978
ISBN 3-540-08476-2

Principles of Magnetic Resonance is a textbook intended for graduate students or others beginning research in magnetic resonance who wish to learn either nuclear magnetic resonance or electron spin resonance. It is intended for physicists, chemists, applied scientists, or others who have had a one year graduate course in quantum mechanics from one of the standard textbooks. The text aims at developing a physical understanding of magnetic resonance and familiarity with the principal theoretical techniques needed to read resonance articles in scientific journals.

Volume 2
O. Madelung, University of Marburg, Marburg, Germany

Introduction to Solid-State Theory

Translated from the German by B. C. Taylor
144 figures. Approx. 450 pages. 1978
ISBN 3-540-08516-5

This book is intended to serve as a textbook in solid-state theory for graduate students of physics and material science. In addition, it should provide the theoretical background needed by physicists doing research in both pure solid-state physics and solid-state physics as applied to electrical engineering.

Volume 3
Z. G. Pinsker, Moscow, USSR

Dynamical Scattering of X-Rays in Crystals

124 figures, 12 tables. Approx. 510 pages. 1978
ISBN 3-540-08564-5

This book presents the first complete treatment of the dynamical scattering of X-rays in perfect and elastically distorted crystals. The theory is systematically developed, experimental methods are discussed, and significant results are illustrated. In comparison to the Russian edition (Nauka 1974) the presentation is substantially enlarged and supplemented by the theory of scattering in elastically distorted crystals and the solution of the multi-beam problem. Reference is made to papers published as recently as 1977.
The book is aimed at scientists and engineers concerned with the technology of single-crystal materials.